Dear Doctor Franklin

Emails to a Founding Father about Science, Medicine and Technology

Stuart A. Green, M.D.

Diane Publishing Company

COLLINGDALE, PENNSYLVANIA

FOR ADRIENNE

A good Wife & Health is a Man's best Wealth

—Poor Richard

Published by
Diane Publishing Company
Collingdale, Pennsylvania
www.dianepublishing.net

ISBN-10: 1-4223-9470-0
ISBN-13: 987-1-4223-9470-0

Sponsored by Friends of Franklin, Inc.
Philadelphia, Pennsylvania
www.friendsoffranklin.org

EMAILS

🗁INTRODUCTION

✉ Emails to Ben Franklin   9

✉ Immers'd in a cask of Madeira wine   13

✉ Better preservation of the English language   17

✉ Taxes we pay for the advantage of long life   19

✉ Closing a long and useful life   21

🗁HEALTH & MEDICAL SCIENCES

✉ Double spectacles   25

✉ Considerable persons   27

✉ To have a flexible catheter   29

✉ A little touch of the gout   33

✉ Disabled by the stone   37

✉ Scarcely escape newspaper defamation   41

✉ Hide not your talents   43

🗁MICROSCOPY & CONTAGEOUS DISEASES

✉ That admirable instrument the microscope   47

✉ Air is replete with infinite multitudes of living creatures   49

✉ An account of the small-pox   51

✉ Suspected to propagate infectious distempers   57

✉ The same distemper is often bred in ships   61

✉ Take one ounce of good Peruvian bark   63

✉ The yellow or bilious fever   67

✉ A continual risque to my health   69

✉ Great numbers of you catch cold   71

✉ A disgrace among the ancient Persians to cough or spit   75

📁ELECTRICITY

✉ Produce something for the common benefit of mankind   81

✉ The electrical matter consists of particles very subtile   85

✉ Secured from the stroke of lightening   87

✉ Mischief by thunder and lightning   91

✉ To the magazines at Purfleet   95

✉ *Erepuit coelo fulmen sceptrumque tyrannis*   97

✉ Electricity in palsies   99

✉ M. Volta's experiment   103

✉ Electricity and magnetism   107

✉ The conducting quality of some kinds of charcoal   111

✉ Some principle yet unknown to us   113

✉ As thro' a vacuum   115

📁CHEMISTRY

✉ You have the philosopher's-stone   119

✉ Wind generated by fermentation   123

✉ Monsieur Geoffroy of the Royal Academy of Sciences   125

✉ This air is not fit for breathing   129

✉ There is nothing unhealthy in the air of woods   131

✉ A candle burned in this air with a remarkably vigorous flame   135

✉ You have set all the philosophers of Europe at work upon fixed air   137

✉ Murdered in mephitic air so many honest harmless mice   139

✉ Method of making salt-petre   141

✉ The power of man over matter   145

✉ On The bad effects of lead   149

📁NUTRITION & DIETARY SCIENCES

✉ I had formerly been a great lover of fish   155

✉ Animal heat arises by or from a kind of fermentation   159

✉ Keep out of the sight of feasts and banquets   161

✉ Eat not to dullness   163

✉ dr f   167

⊠ Do not make an expensive feasting 169

⊠ White as your lovely bosom 171

🗁 EXERCISE SCIENCE & PHYSIOLOGY

⊠ Running, leaping, wrestling and swimming 175

⊠ *The Cause of Heat of the Blood* 179

⊠ The globules of the blood 183

⊠ The frame of the auricles or ventricles of the heart 189

🗁 GEOLOGY, EARTH SCIENCE & EVOLUTION

⊠ After the earthquake, an unusual redness appeared in the western sky ... 195

⊠ A purse made of the asbestos 199

⊠ One continent becomes old, another rises into youth and perfection ... 201

⊠ The surface of the globe would be a shell 205

⊠ With strata in various positions 211

⊠ The nature of this globe 215

⊠ The prolific nature of plants or animals 219

🗁 ATMOSPHERIC SCIENCE & TECHNOLOGY

⊠ *Suppositions and Conjectures on the Aurora Borealis* 225

⊠ The balloon we now inhabit 227

⊠ Thermometers are often badly made 233

⊠ The dissolved water in the air 237

⊠ As easy as pissing abed 239

⊠ Whirlwinds at land, waterspouts at sea 243

⊠ Some curious and useful discoveries in meteorology 245

⊠ Different degrees of heat 249

⊠ The author's description of his Pennsylvania fire-place 251

⊠ I soon became a thorough Deist 253

⊠ A fool and a madman 255

🗀MARINE SCIENCE AND OCEANOGRAPHY

✉ The edges of the Gulph Stream 259

✉ Divide the hold of a great ship into a number of separate chambers 261

✉ One of Cook's vessels 267

✉ Draw a boat in deep water 271

✉ Their waters are fresh quite to the sea 273

✉ The effect of oil on water 275

✉ Propelling boats on the water, by power of steam 279

✉ A sooty crust on the bottom of the boiler 283

✉ A large boat rowed by the force of steam 287

✉ The types he had been reading all his life 291

🗀PSYCHOLOGY & SOCIAL SCIENCE

✉ Varying degrees of imagination 295

✉ A convenient and handsome building 301

✉ A fine house…half a mile from Paris 305

✉ A good war or a bad peace 307

✉ Healthy and wealthy and wise 311

✉ May be bound in one volume…with a compleat index 313

✉ Disappointment and fallacious hope 319

INTRODUCTION

✉ Emails to Ben Franklin   9

✉ Immers'd in a cask of Madeira wine   13

✉ Better preservation of the English language   17

✉ Taxes we pay for the advantage of long life   19

✉ Closing a long and useful life   21

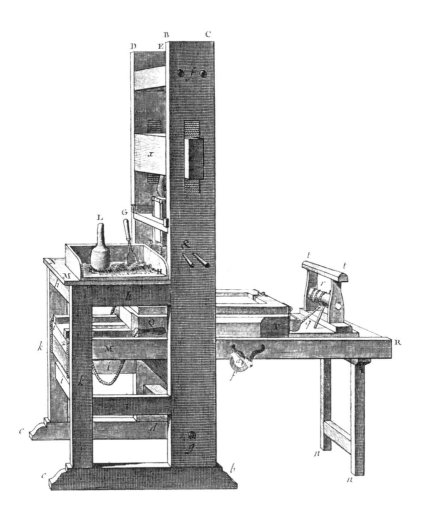

18ᵗʰ century printing press

From: "Stuart Green" <stuartgreenmd@yahoo.com>
To: "Roy Goodman" <rgoodman@amphilsoc.org>
Subject: **Emails to Ben Franklin**

Hi Roy:

Congrats on your re-election as president of *Friends of Franklin*. I'm sure you'll continue doing a great job.

I've attached to this email a document you might find interesting, and want to share with other members of F of F. Not long ago, I read that the family of baseball legend Ted Williams had his head severed from his body—after he died, of course—and frozen in liquid nitrogen. The process will allow The Thumper future resurrection when medical science solves the obvious problems.

When I checked the websites of companies doing this sort of thing, I found that they admire Ben Franklin as a patron saint, based on his letter to Barbeu-DuBourg. You must remember it: Franklin mentioned watching flies that had drowned in wine revive in the sun and suggested he'd like to do the same himself and come back to life in the future.

I concluded that perhaps Franklin did, indeed, have himself preserved like those drowned flies and will soon be dug up during construction near Franklin Court in Philadelphia.

Since Franklin knew so many of the key people advancing science and technology during his lifetime, I thought he'd like to know what's happened since his era, especially in those disciplines that held his interest. So I decided to write Franklin a series of emails covering the sciences that most engaged him while alive. That way, if and when he's revived, he'll have a ready file of background information.

I also assumed memory problems at his advanced age, so I reviewed for him the circumstances surrounding his involvement with scientific, technical and medical matters. Likewise, since he might not remember the many individuals in his life saga, I created portraits of his colleagues and other science luminaries, based on engravings and other contemporary images. I inserted them into the emails as I wrote, adding illustrations from 18th century encyclopedias of science in my own library, especially Middleton's 1777 *Dictionary of Arts and Sciences* and Humphreys' 7 volume *Nature Display'd* (1750). As Curator of Printed Materials at the American Philosophical Society's Library, you might know these books—both borrowing heavily from the great French encyclopedias of that era.

Needless to say, I also used images from Franklin's own work and from several classics of science, including Lyell's *Principles of Geology*, Darwin's *The Descent of Man*, and others as well. I even procured a drawing from a surgery textbook I wrote more than two decades ago. Additionally, I made new illustrations to help explain certain modern principles of science, converted a few of my own photographs into line-art, and downloaded pictures from NASA's website—all for Benjamin Franklin's edification

To ease Franklin into the project, I typed my first email in the 18th century manner of writing, but got out of that mode as soon as I could.

I've stuck to topics I know most about: medicine, science and technology. I've held off writing about numerous other topics of equal importance to Franklin—agriculture, slavery, finance, government, and so forth—except when summarizing historical events in the context of industrial and technological development.

You know, Roy, I wouldn't have gotten involved in this project but for a proposal by Ellen Cohn, editor-in-chief of the multi-volume *Papers of Benjamin Franklin*. She and her predecessors and colleagues at Yale's University Library have done a remarkable job assembling everything Franklin wrote during his lifetime and everything written to him, and publishing that material with well-researched annotations.

When the Packard Humanities Institute of Los Altos, California, digitized and made all of Franklin's papers available to the public—first on a CD and now at a website—they performed a significant service to scholarship.

Do you ever visit the Yale site (www.yale.edu/franklinpapers/index.html) or look through the Franklin papers themselves (franklinpapers.org/franklin/)? I sure did for this project.

I don't know if I ever told you this, but I got interested in Benjamin Franklin in a most peculiar way. As an academic orthopaedic surgeon, I researched the subject of sham (placebo) surgery for pain relief and Googled to a website of the Royal College of Physicians of Edinburgh. There I learned to my surprise that Benjamin Franklin helped design the first placebo-controlled medical research.

Like most Americans, I viewed Franklin as the witty and clever Founding Father who devised the lightning rod and enjoyed the company of ladies. I didn't realize how many diverse scientific disciplines honor Franklin's contributions to their respective fields, including particle physics, oceanography, meteorology, toxicology, earth sciences, economics, population dynamics, metabolism, plant ecology, and epidemiology. Printers, swimmers, songwriters, chess players, musicians, Masons and *bon vivants* also claim Franklin as their own. Even the deep thinkers get into the act: Franklin is acknowledged as the father of our native-born school of philosophy, American Pragmatism

I later understood that Franklin's political leverage at home and abroad followed the acclaim for his scientific discoveries and the 1751 publication of his book, *Experiments and Observations on Electricity made at Philadelphia in America*, which, like the attached emails, is a series of letters about science.

I soon became a collector of Frankliniana and joined the *Friends of Franklin* where I met others who shared my fascination with our nation's great patriarch.

During a Friends of Franklin outing in Annapolis, Maryland a few years ago, Ellen Cohn suggested I write a book about Franklin's interests in medical and scientific subjects. After thinking about her proposal and considering the Barbeu-DuBourg letter, I asked myself: Why write *about* Franklin when I should be writing *to* Franklin; hence, my emails.

Franklin lived, of course, during the Enlightenment, the extraordinary epoch in human history that spawned, among other things, the scientific revolution.

I agree with those historians who ascribe the Enlightenment's intellectual foundation to Sir Francis Bacon, counsel to Queen Elizabeth and later Lord Chancellor during the reign of James I. It's too bad Bacon got sacked for accepting bribes while in office, although he later disavowed his "confession."

Francis Bacon

Bacon penned his great 1620 book *Novum Organum* to overthrow the Aristotelian method of deductive reasoning—we now call it armchair philosophizing—and substituted instead inductive analysis, always starting with observed facts.

Bacon's process of inquiry proved particularly valuable in the prism and optical research of Isaac Newton in 1666.

Franklin, as you know, was familiar with Newton's *Opticks* although he lacked the math and Latin background to tackle Newton's masterwork *The Mathematical Principles of Natural Philosophy*. Furthermore, I suspect that the book's subject matter—orbital physics of planets—didn't hold much interest for Franklin, intent as he was about earthly matters.

By the time Franklin began experimenting with electricity in 1747, the scientific method was well established and rational thought had replaced superstition and reliance on Scripture in the minds of many researchers (although most assumed they were gaining insight into the Creator's construction principles with their research). Before Franklin's electricity experiments, however, no researcher of the late 17th or early 18th centuries had come up with anything really practical to show for their effort. I mean, Roy, how much does a farmer in the field or a merchant in his shop care about the orbit of Mercury or the speed of falling balls?

Isaac Newton

Franklin, however, came up with a practical use for his electricity discoveries, something that not only protected people and property but also had deep theological implications. After all, lightning was God's precision punishment, and Franklin diminished that power with his rod.

With fame and credibility from his scientific endeavours, Franklin became a political being. So much so that as historian Joseph Ellis once put it, if you took a snapshot at every important event in early American history, Benjamin Franklin would be in the picture.

Not long ago, I solicited Stan Finger's opinion about my BF emails. (His wonderful book *Dr. Franklin's Medicine* served as inspiration for this project.) Stan made helpful suggestions but also mentioned that Franklin wouldn't have survived emersion in alcohol—a substance that destroys nerve tissue. He thus tried to convince me that my emails wouldn't ever reach their intended target.

Rather than wait for something that might never happen, I've decided to send the emails to Friends of Franklin for distribution to its members and anyone else interested in Benjamin Franklin's noteworthy involvement in science, medicine and technology. I'll post citations and footnotes on the friendsoffranklin.org website. I hope you and the membership like reading what I've written to Ben Franklin as much as I enjoyed communicating with him.

Stan, by the way, hopes forensic geneticists will someday exhume Franklin's body for DNA testing. If and when that happens, Stan will get a chance to test his hypothesis that poisoning from lead-salted wines—common in Europe during Franklin's lifetime—damaged the great man's kidneys, resulting in both gout and renal stones, and secondarily caused his lung abscess and death. Won't Stan (and everyone else) be surprised if the body entombed at 4th and Arch doesn't match the DNA of Benjamin Franklin's descendents?

From: "Stuart Green" <stuartgreenmd@yahoo.com>
To: "Benjamin Franklin" <dr_benjamin_franklin@yahoo.com>
Subject: **Immers'd in a cask of Madeira wine**

Dear Doctor Franklin:

Please forgive my audacity for writing this Communication tho' you have been suppos'd dead more than two hundred years. I have faith you will read it and have many things to ponder thereby. I write in your own Style to ease Difficulty of Discernment, for much has chang'd in our Mutual Language, whilst many old Words are yet retain'd. Future Communications shall, to my ear, have a more Modern Tone.

My name is Stuart Green and I am a Physician much interest'd in your Life and Times. I happily reside in the United States of America, a Nation today "populous and powerful" as you had predict'd so many years ago.

Our Country recently celebrated the three hundredth Anniversary of your Birth, for you are as admir'd by Posterity as by your own Acquaintances. Indeed, your Fame has increas'd over time: you are now rever'd as a great Sage and Patriarch of our Country.

Your Progeny, descend'd from your daughter Sarah's children, are as numerous as the Cobblestones on Market Street. Metallic Rods, from your time 'til now, protect every manner of Building from Lightning strikes.

While studying the Facts of your illustrious Life I read a Newspaper report of your Demise written by Dr. John Jones who attend'd you in your presum'd final Hours. The Particulars of the Account were, it seemed to me, more compleat than necessary for Person of your Fame. I wonder'd why Doctor Jones, would write such a detailed Depiction, except to persuade the Public of its reality tho' untrue.

I puzzled on this matter 'til I chanc'd on this Letter you wrote to Jacque Barbeu-DuBourg:

I have seen an instance of common flies…drown'd in Madeira wine…Having heard it remark'd that drowned flies were capable of being reviv'd by the rays of the sun, I proposed making the experiment upon these; they were therefore expos'd to the sun…In less than three hours, two of them began by degrees to recover life…and soon after began to fly, finding themselves in Old England, without knowing how they came thither.

I wish it were possible…to invent a method of embalming drown'd persons, in such a manner that they may be recall'd to life at any period, however distant. For having a very ardent desire to see and observe the state of America a hundred years hence, I should prefer to any ordinary death, being immers'd in a cask of Madeira wine...to be later recall'd to life by the solar warmth of my dear country!"

Upon further conjecture, I reasoned the following: In March of 1790, a Month before your record'd Death, you and Dr. Jones conspir'd to repeat your Madeira'd fly experiment but with a living Person—yourself—instead of an Insect so encask'd. As you neared Death from a ruptured Lung Abscess and suffer'd exceedingly from your Gout and Bladder stones Doctor Jones, tho' perhaps reluctant, agreed with your final Scheme.

And thus before you breath'd your last, while in delirium from Opium, Doctor Jones (helped by your grandsons Benjamin and Temple) place you in a large oaken Barrel. Rapidly filling the remaining space with sweet Madeira, Doctor Jones thereby preserv'd for Posterity the greatest Philosopher of his age.

At the end of the Deed, Doctor Jones pound'd home a thick oak Lid upon which was carved: "Open not this Cask 'til be Found for Stone and Gout a Cure. B Franklin, Printer."

But, as *Poor Richard* said: "When the Wine enters, out goes the Truth." And so Jones, needing a portly Corpse, stole from Pennsylvania Hospital the Body of a Man recently deceased and nameless to the Stewards there.

The unknown's Proxy's Funeral was grand indeed; 20,000 attended, it is said. The Pallbearers may have shoulder'd a foreign pauper treated in death like a Prince of State. He perchance lays today repos'd next to your dear Deborah in the Graveyard of Christ Church, Philadelphia.

In the meantime, Doctor Jones and your grandsons, sworn to secrecy, in dark of Night, I imagine, buried your Barrel in the Common Ground of your beloved town.

At present, there is much digging in Philadelphia's center, so I assume workers will soon enough strike their Shovels against a strange oak Cask. The words on the Lid will doubtless cause great Commotion thru the land. Your Descendants, upon consultation, will no doubt permit you revitaliz'd by electric Jolt.

Upon return to Life, you will quickly learn that we still employ Extracts of Opium for Pain relief and the Delirium caused thereby is just the same. Luckily tho' both Gout and Stones are now easily cur'd. I shall explain how soon enough.

I have reason'd that you will, upon regaining your Senses, be much bewilder'd by all you see. I, therefore, believing I have correctly surmised both your false Interment and your future Revival, undertake the difficult but worthy Task of informing you about interesting Things that have transpir'd since your *mort-faux*.

You will have much Time to read while you recover from Encaskment. To fill those Hours, I will prepare many Communications focusing attention on those Things of interest to you during your Lifetime.

I fear that by spending more than two hundred Years immersed in Madeira, your remarkable Brain has become besotted and your Memory diminished. I thought you might find benefit in being reminded of your numerous Accomplishments during your long and prodigious Life, to better discourse on these Matters when queried.

I will write of your Friends now esteem'd for great and lasting achievements. I shall report to you upon a favorite Subject of yours, the advancements in Science and in Medicine that in life command'd your attention. For example, consider this: The six week journey from Philadelphia to London is now accomplish'd in six hours; Small-Pox is stricken from the List of afflictions; and Men have walk'd upon the Moon and returned safely from thence hither.

To explain how these things came to pass, I must remind you of your ideas about Electric Storms and Waterspouts, about the shape of Rocks and Toads in Stone and many more things. You will learn from me which of your many Conjectures are now consider'd correct and which few, wrong-head'd.

To describe such Changes, I must provide you a new Vocabulary, for many Words have been lately added to our belov'd English. Clearly, you have much to absorb that will demand your patient Attention. The Language of these communications will evolve, as do the Ideas, so at the end of them both, I will write and you will read as a modern Person.

I choose to write not by Quill and Paper but by a prevailing manner called Electronic Mail, often named "email" by the many that employ it. I shall spend considerable time explaining this as it is mostly deriv'd from your electrical Experiments.

Do I err upon who lies beneath the Franklin churchyard Stone? I hope not. The good Doctor Jones himself succumb'd a year after your alleged demise. Bennie perish'd of The Fever seven years thereafter, and Temple, the last of my presum'd Conspirators to pass away, like the other two, never said a word that would support my curious thoughts. Tho' *Poor Richard* said, "Three can keep a secret if two of them are dead," I reply, "Three best keep a secret if all of them are dead!"

And so, Sir, as I wait your Cask's unearthing, I assume the rightness of my Notions and prepare for you this History of your Ideas, for that will be the Subject henceforth herein.

For my Effort, I pray to never hear Poor Richard's words, "Who has deceiv'd thee so oft as thyself?"

PLAIN CONCISE

PRACTICAL REMARKS

ON THE TREATMENT OF

WOUNDS AND FRACTURES;

TO WHICH IS ADDED, A SHORT

APPENDIX

ON

CAMP AND MILITARY HOSPITALS;

PRINCIPALLY

Defgned for the Ufe of young MILITARY SURGEONS,

in NORTH-AMERICA.

By JOHN JONES, M. D.

Proeffor of Surgery in King's College, New-York.

NEW-YORK:

by JOHN HOLT, in Water-Street, near the

Coffee-Houfe.

M,DCC,LXXV.

First American book on fracture care, written by Dr. John Jones.

From: "Stuart Green" <stuartgreenmd@yahoo.com>
To: "Benjamin Franklin" <dr_benjamin_franklin@yahoo.com>
Subject: **Better preservation of the English language**

Dear Doctor Franklin:

Several days have pass'd since I last email'd you about your Present Situation, assuming that I have correctly surmis'd it. I must here inform you of changes in English usage so you might modernize rapidly your Discourses and Communications. English writers no longer capitalize Nouns in the middle of Sentences in the manner of German; therefore I will not do so hereafter.

We do not shorten the past tense of verbs by inserting an apostrophe in place of the last "e" in the word. Thus, we presently write *passed* instead of *pass'd*. I shall follow the modern scheme henceforth.

Do not assume, however, that the shortening apostrophe is gone from our writing, for it is not. We still write and say *can't* for *cannot*, *I'll* for *I will*, *won't* or *wouldn't* for *would not*, &c, which, by the way, we now write as *etc.*

To help modernize your communication, in future emails I'll transform your own quoted words to reflect these changes in our language. What you wrote as "oblig'd to be on the Water" I shall record as "obliged to be on the water."

In our written and printed pages, the tall *s* no longer appears, for it too closely resembles the letter *f* for easy comprehension. (Children, when first seeing the handwritten version of the Declaration of Independence, often ask their teachers the meaning of "in Congreff Affembled.")

Endeavour is now spelled *endeavor* on this side of the water. This change has occurred in America but not in Great Britain, where the words *color* and *honor* are still written as *colour* and *honour*—often the only way to tell if a book was printed in Philadelphia or London.)

The rhythm of the language has also changed. We use short verbs and adverbs whenever possible. Our salutations are much less formal. Nobody alive today would end a communication *Your Most Humble and Obedient Servant*. We complete our letters instead with *sincerely yours*, or *cordially yours*, or *truly yours*, even after writing something insincere, not cordial, or untrue.

I can't think of additional changes. The rest will become obvious as you read my emails.

I know that in your former life, you expressed concern about the use of foreign languages on American soil. In 1754, you and other Pennsylvania delegates suggested, "That for the better preservation of the English language…it may be proper to require by a law that all written contracts be in English and to give some encouragement to English schools…."

English has remained the common language of Americans, although immigrant groups continue to speak their native tongues at home. They also prefer to shop and do business with those sharing their birth language. The first generation of native-born children, however, learns English rapidly, often serving as interpreter for their parents.

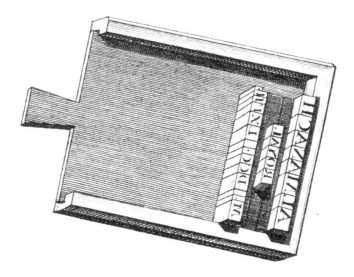

Forgive me in advance, Dr. Franklin, if I quote from your own documents, books, and letters. While your final letters are as lucid as your earliest ones, the fullness of your life, the variety of your interests and the astonishing number of your communications still extant, suggests that you couldn't possibly recall all you've written. I'll therefore remind you about what you wrote, and to whom.

Even though you received doctorate degrees from both St. Andrew's University in Scotland and Oxford University, some people might object to addressing you as "Doctor Franklin" since both degrees were honorary. I'll inform such naysayers, however, that your diplomas were conferred for important scientific discoveries in electricity. Thus, you earned the designation Doctor used during your lifetime.

Altogether more than 400 books have been written about your life. You'll no doubt have many belly laughs reading what some "authorities" have said about you. Don't criticize them too harshly—the passage of time fogs accurate hindsight. Each historian must shine light on a subject and interpret the reflected luminescence as he or she sees fit.

From: "Stuart Green" <stuartgreenmd@yahoo.com>
To: "Benjamin Franklin" <dr_benjamin_franklin@yahoo.com>
Subject: **Taxes we pay for the advantage of long life**

Dear Doctor Franklin:

My father, a Bostonian like you, was 91 at his passing. He developed memory problems in his latter years, often confusing the sequence of events in his earlier life. Fearing you might do the same, I've decided to provide you with a brief chronology of your life, since my emails won't follow an historical sequence. Instead, they'll track your interests and ideas through time—from yours to now.

I've divided your life into phases based on your residence at the time. I suspect you'll take exception to this grouping, but for now, follow along.

(We call this kind of catalogue a *bullet list*.)

- **BOSTON, AGE 0-17:** Began life in 1706. Continued through two years of formal schooling and early years working in your father Josiah's soap and tallow business. He later apprenticed you to brother James, whose harsh tutelage at his print shop you escaped at age sixteen.
- **PHILADELPHIA, AGE 17-18:** Worked as a printer's assistant. Boarded in home of future wife Deborah Reed.
- **LONDON, AGE 18-20:** Unsuccessfully attempted to buy printing press and type but a promised Letter of Credit from Pennsylvania's Royal Governor never arrived. Worked as printer to earn return passage to America.
- **PHILADELPHIA, AGE 21-51:** Gradually became a successful printer (*Pennsylvania Gazette, Poor Richard's Almanack*) businessman and public person. Married DR after having illegitimate son William with another woman. Had daughter Sarah and son Francis (who died of smallpox at age 4). Retired at age 42 to conduct electric experiments. Made important discoveries (including lightning rod) at age 43. Began to win awards and honorary degrees. As colonial postmaster, attempted but failed to unite colonies for administrative purposes (Albany Plan).
- **LONDON, AGES 51-57:** Represented Pennsylvania as political and business agent in Europe. Made many London friends. Received two honorary doctorate degrees (St. Andrew's, Oxford) for experimental work.
- **PHILADELPHIA, AGE 58:** Tours New Jersey, New York and New England inspecting post offices. Visits charity school and develops "a higher opinion of the natural capacities of the black race than I had ever before entertained."

- **LONDON, AGES 59-69:** Helped get Stamp Act repealed. Renewed acquaintances with British men of science and letters. Represented four colonies (Pennsylvania, Georgia, New Jersey, Massachusetts). Became increasingly bitter towards British government's colonial tax policies. Left England in anger after being accused of stealing and disseminating private anti-colonist letters.

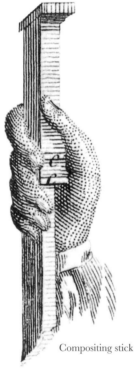

Compositing stick

- **PHILADELPHIA, AGE 70:** Helped draft Declaration of Independence. Traveled north to unsuccessfully plea with Canadians to join rebellion against Mother Country.
- **PARIS, AGE 70-79:** Raised money and secured arms, supplies, and gunpowder for American Revolution. Convinced Louis XVI to sign treaty supporting U.S. cause. Later negotiated peace with Great Britain and obtained other international treaties as well (Sweden, Morocco).
- **PHILADELPHIA, AGE 80-84:** Became President of Pennsylvania and later host of, and oldest delegate to, the Constitutional Convention. Became president of Pennsylvania abolitionist society. Planned experiment with faked death and funeral. Hidden by Dr. Jones in Madeira filled-barrel somewhere in Philadelphia.

My father told me before he passed away that a long life, while desirable, means attending many funerals for friends and relations. You however put it best when you wrote: "Loss of friends and near and dear relations is one of the taxes we pay for the advantage of long life, and heavy tax indeed it is!"

From: "Stuart Green" <stuartgreenmd@yahoo.com>
To: "Benjamin Franklin" <dr_benjamin_franklin@yahoo.com>
Subject: **Closing a long and useful life**

Dear Doctor Franklin:

As your health improves, you'll chuckle over your own death announcement. I've transcribed it for your amusement. This version appeared in John Fenno's *Gazette of the United States* within a few days of your encaskment. To me, it seemed too detailed; I suspect that the author, Dr. John Jones, tried to persuade the public of its truthfulness, rather than be ridiculed for participating in your Madeira project. Here it is:

> PHILADELPHIA, April 20, 1790
> Died on Saturday night, in the 85th year of his age, the illustrious BENJAMIN FRANKLIN, of this city. His remains will be interred tomorrow afternoon, at four o'clock, in Christ Church burial ground.
>
> We are favored with the following short account of Doctor Franklin's last illness, by his attending Physician.
>
> THE stone, with which he has been afflicted for several years, had for the last twelve months, confined him chiefly to his bed; and during the extreme painful paroxysms, he was obliged to take large doses of laudanum to mitigate the tortures—still, in the intervals of pain, he not only amused himself with reading and conversing cheerfully with his family and a few friends, who visited him, but was often employed in doing business of a public as well as private nature, with various persons, who waited on him for that purpose; and in every instance displayed, not only that readiness and disposition of doing good, which was the distinguishing characteristic of his life, but the fullest and clearest possession of his uncommon mental abilities; and not frequently indulged himself in those *jeux d'espirits* and entertaining anecdotes, which were the delight of all who heard him.
>
> About sixteen days before his death, he was seized with a feverish indisposition, without any particular symptoms attending it till the third or fourth day, when he complained of pain in the left breast, which increased until it became extremely acute, attended with a cough and laborious breathing. During this state, when the severity of his pains sometimes drew forth a groan of complaint, he would observe—that he was afraid he did not bear them as he

ought—acknowledged the grateful sense of the many blessings he had received from that Supreme Being, who had raised him, from small and low beginnings, to such high rank and consideration among men—and made no doubt his present afflictions were kindly intended to wean him from a world, in which he was no longer fit to act the part assigned to him.

In this frame of body and mind he continued till five days before his death, when his pain and difficulty of breathing entirely left him, and his family were flattering themselves with the hopes of his recovery, when an imposthumation, which had formed itself in his lungs, suddenly burst, and discharged a great quantity of matter, which he continued to throw up while he had sufficient strength to do it, but, as that failed the organs of respiration became gradually oppressed—a calm lethargic state succeeded—and on the 17th, at about 11 o'clock at night he quietly expired, closing a long and useful life of 84 years and three months.

It may not be amiss to add to the above account that Dr. Franklin, in the year 1735, had a severe pleurisy, which determined in an abscess of the left lobe of his lungs, and he almost suffocated with the quantity and suddenness of the discharge. A second attack of a similar nature happened some years after this, from which he soon recovered, and did not appear to suffer any inconvenience in his respiration from these diseases.

Gazette of the United States.

PUBLISHED WEDNESDAYS AND SATURDAYS BY *JOHN FENNO*, No. 9, MAIDEN-LANE, NEW-YORK.

[No. 108.—Vol. II.]   SATURDAY, *APRIL* 24, 1790.   PRICE *THREE* DOLLARS PR. ANN.

Masthead of John Fenno's newspaper

You can understand, I hope, my skepticism about your alleged final moments. It stuck me that no doctor would provide so many details about a death unless he was determined to embellish an untruth with vivid particulars to increase its veracity. (If I'm wrong about this, Sir, I'm on a fool's errand, writing to a dead person!)

Imposthumation, by the way, is a word we no longer use. We'd say empyema (also employed in your time) to describe your lung condition. I'll tell you more about the cause of empyemas in a later email.

I hope you can easily read what I've written; I presume Dr. Jones left your double spectacles on your nose before encaskment. Since your eyeglasses will be the first item you see on revival, I'll tell you about eyewear development in my next email.

HEALTH & MEDICAL SCIENCE

✉ Double spectacles 25

✉ Considerable persons 27

✉ To have a flexible catheter 29

✉ A little touch of the gout 33

✉ Disabled by the stone 37

✉ Scarcely escape newspaper defamation 41

✉ Hide not your talents 43

How I assume you'll look upon revival.

From: "Stuart Green" <stuartgreenmd@yahoo.com>
To: "Benjamin Franklin" <dr_benjamin_franklin@yahoo.com>
Subject: **Double spectacles**

Dear Doctor Franklin:

No double spectacles have ever been found among your possessions or for that matter, any ordinary eyeglasses either. If Dr. Jones failed to place spectacles in your Madeira cask, the hospital's doctors will have a pair made when you come to your senses.

You're universally recognized as the inventor of *bifocal* eyeglasses (you called them "double spectacles"), based on your 1784 letters to British philanthropist George Whatley. (Recall writing to Whatley from your Passy residence near Paris, while serving as our nation's envoy to the Court of Louis XVI.) You mentioned in one letter that you formerly carried two pairs of spectacles while traveling, one for reading and one for distance viewing, continually exchanging them as you alternately read and observed the passing scenery. To eliminate this nuisance, you had eyeglasses made with both distance and reading lenses in the same frame.

Most people assume that you invented bifocals during your years in France to enable you to both see the food on your plate and "the faces of those on the other side of the table," because, as you put it, "when one's ears are not well accustomed to the sounds of a language, a sight of the movements in the features of him that speaks helps to explain."

Some historians have suggested that you first devised bifocals in the late 1730s—around the time you began advertising eyeglasses in your *Pennsylvania Gazette*—at least fifty years before you went to France during the American Revolution. No doubt, you'll settle this question.

In April 1779, I hope you'll recollect, you received a letter from an eyeglass-maker complaining about making double spectacles to your order. He damaged three lenses while trying to cut them in half. Moreover, he wasn't the only one having difficulties; opticians couldn't stop the half-lenses from falling out of the frames. Another nuisance—the circular lenses rotating in the rims—obliqued the seam between the two halves. (Thomas Jefferson solved that problem by setting the two half lenses in *oval* frames.)

News of your bifocals invention spread quickly after your encaskment. In 1790, the year of your pronounced demise, Dr. Thomas Rawley published a book on diseases of the eyes and eyelids. In his chapter on spectacles, Rawley quoted your 1785 letter to George Whatley and said that he recommended double spectacles to friends, who were delighted with them. A year later, the public first learned of your double spectacles in *Massachusetts Magazine*.

As your bifocals became popular, wearers complained about dirt collecting in the seam between the half lenses, causing a visible black line. This seemingly trivial irritant proved a driving force behind the evolution of bifocals.

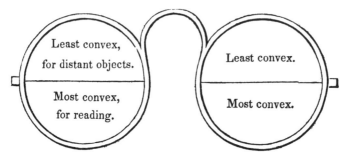

Your own diagram of double spectacles

Many competing designs featured ways to minimize the seam. The best of these, Dr. Franklin, was the Ultex lens, a single piece of glass, thin on the top for distance vision and thicker on the bottom for reading. Trifocal eyeglasses first appeared in the early decades of the 20th century. Many other lens designs followed.

During the 1940s, as the world went through a global war, British chemists developed transparent unbreakable substances called *acrylic plastics*, now the material of modern eyeglass lenses.

By the mid-20th century, the only remaining problem with bifocal and trifocal glasses was the sudden transition between different lens sections, an effect that no doubt annoyed you as well.

The final refinement of your double spectacles eliminated the sharp transition. *Progressive lenses* have a smooth change between regions of different focal length. They must be custom made for each wearer, based on the inter-pupils distance. Poorly made progressive eyeglasses cause dizziness and nausea. Even with a good fit, some people never get accustomed to progressives no matter how hard they try.

For many of us, however, progressive lenses have miraculously restored the perfection of youthful vision to aging eyes. Indeed, I'm reminded of what you wrote Whatley: you were "happy in the invention of double spectacles, which serving distant objects as well as near ones, make my eyes as useful to me as ever they were." I couldn't agree more.

By the way, I hope Dr. Jones folded your fine green suit into your barrel along with your double spectacles. I often envision you sitting at a table, dressed in that outfit, reading my emails with your chin resting on your thumb, just as depicted in David Martin's 1767 portrait. I know you liked that painting so much you ordered a replica made, now owned by the American Philosophical Society. I've seen it there myself. But for that inspiring image, continuing these emails might prove difficult indeed.

From: "Stuart Green" <stuartgreenmd@yahoo.com>
To: "Benjamin Franklin" <dr_benjamin_franklin@yahoo.com>
Subject: **Considerable persons**

Dear Doctor Franklin:

As news of your revival spreads, many individuals will clamor to meet you. Your descendents, now numbering in the thousands, will doubtless profess filial devotion. Likewise, many persons surnamed Franklin may claim descent from you, but as all your progeny came through your daughter Sarah Bache, the Franklin name must have come from someone else's descendents. For these reasons, I suggest you decline meeting with family members until you've fully regained your senses and refreshed your memory with my emails.

Certain individuals who in your day were called "considerable persons" will insist upon making your acquaintance as soon as you revitalize—if only to later brag about the encounter. I suggest, Sir, you limit such meetings to those individuals who owe their position to your own activities more than two centuries ago. As you will likely follow protocol, I recommend that you greet them in order of their importance or influence, with the President of the United States first.

I needn't remind you of your proposal for a chief executive of Britain's North America colonies in your 1754 *Plan of Union*. Our nation's current leader is, in many ways, your wished-for "President-General." You repeated the President-General idea again in your draft for an Articles of Confederation, submitted at the Second Continental Congress in 1775. As you know, the Articles eventually agreed upon in 1777 created a toothless central government with no power over individual states. Your advocacy for a predominant federal authority, however, influenced the U.S. Constitution, still the nation's governing document.

Although your encaskment occurred during the first of Washington's eight years in office, you knew six American presidents personally. After Washington, John Adams assumed the leader's post, followed by Thomas Jefferson. He doubled the size of the United States in 1803 by purchasing all of France's North American holdings. James Madison and James Monroe—Virginia delegates to the Constitutional Convention—served as our fourth and fifth presidents.

You'll never guess who became our sixth president: young Quincy, John Adam's son. Recall that he spent time in Paris with your grandson Temple while you, his father and John Jay negotiated peace with England. Quincy proved quite a good president, which shouldn't surprise you, considering how bright he was. Until recently, the Adamses were the only father and son who served as American presidents, but the man you'll likely encounter is also a former president's son.

Only men of European ancestry have occupied the president's office. We've had only one Catholic president—no Jews, Blacks, Native Americans, Asians or women. Deranged assassins slew four presidents in office, which explains the security arrangements that'll precede the President's visit.

Out of courtesy you should meet with America's Chief Justice along with the eight Associate Justices of the Supreme Court. I recommend, however, that you deflect their questions about specific issues. Today's legal debates focus on what the *Founders* meant by freedom of speech, religion or the press. The justices will want answers from you because these matters are much on their minds.

Pennsylvania's Governor should be next because you functioned as that state's chief executive twice in your career. Both he and the Mayor of Philadelphia will advise you of developments in your adopted state. As for your birthplace, Boston's Mayor and the Massachusetts Governor also deserve a short visit.

Seventy men have served as Postmaster General of the United States since you occupied that position. Posterity credits you with founding an efficient postal enterprise that served the nation rather well for more than 150 years. By the 1960s, however, the postal system became hopelessly old-fashioned, failing to take advantage of new technology. Congress and the President converted our Postal Service to a profit-making corporation wholly owned by the federal government. The present postmaster will fill you in on the details of modern postal operations. Permit extra time for the visit.

Historians claim that your support of the Great Compromise—establishing equal states' representation in the Senate and proportional representation in the House of Representatives—was essential for ratification of the Constitution. For this reason, members of both houses of Congress will request an audience. If you don't restrain them, all 100 senators and 438 congressmen and congresswomen will offer compelling reasons why they—rather than anyone else—deserve time with you. Our political system, I regret to inform you, became partisan shortly after you died. Whigs and Tories of England have become the Democrats and Republicans of our times. On many issues, the parties oppose each other for the sake of dissent itself. For this reason, I recommend meeting with a single leader from each chamber and then only briefly, lest one or the other try to solicit you to their side.

You also should parlay with certain officials—the Secretary of State and the Ambassadors to France, England, Sweden and Morocco—whom you'll query about the favorable outcome your foreign treaties of peace, amity and commerce.

My next email will return to health matters to consider your problems voiding.

From: "Stuart Green" <stuartgreenmd@yahoo.com>
To: "Benjamin Franklin" <dr_benjamin_franklin@yahoo.com>
Subject: **To have a flexible catheter**

Dear Doctor Franklin:

I suspect that after asking for your spectacles, you'll want to void. Until surgeons remove the bladder stone that caused difficulty as you grew older, you'll have as much obstruction now as you did before encaskment. (It's said you sometimes stood on your head to dislodge the stone.)

The first doctor you'll encounter for your voiding problem will approach you with a flexible tan-colored catheter and announce that *you* invented it, even though you never saw such a thing while alive. It's often said that Benjamin Franklin invented the flexible urinary catheter, so don't act surprised. You'll readily admit, however, that the modern "Foley" catheter doesn't resemble the device you proposed for your brother John's obstruction.

We have today no record of John's request, just your reply of December 8, 1752, the only place you described a catheter. You wrote, "Reflecting yesterday on your desire to have a flexible catheter, a thought struck into my mind how one might possibly be made." You "went immediately to the silversmith's, and gave directions for making one, (sitting by 'till it was finished)."

As I understand it, Dr. Franklin, your catheter consisted of a thin strip of silver wound around a thick wire core with each layer of silver partially overlapping the one below.) When complete, the silversmith could remove the wire, leaving a hollow tube similar to what we now call flexible conduit tubing. To finish the assembly, you had small "pipes" braised to both ends of the silver spiral.

Obviously, such a device would leak between the layers, so you advised John to either cover the catheter with "small fine gut" soaked in alum or, alternatively, use "a little tallow rubbed over it, to smooth it and fill the joints."

Recall that you informed John the device was "as flexible as can be expected in a thing of the kind" and that it should "readily comply with the turns of the passage, yet has stiffness enough to be protruded."

The next thing you told your brother was quite remarkable: You said that if the catheter didn't have enough stiffness to advance properly, "the enclosed wire may be used to stiffen the hinder part of the pipe while the fore part is pushed forward; and as it proceeds the wire may be gradually withdrawn."

This technique, using a *stylet* to stiffen a flexible catheter, is familiar to anyone today who has had a blood vessel catheterized. Doctors and *nurses*—the name we give to hospital sisters—are particularly adept at the maneuver that advances the flexible plastic catheter into a vein while simultaneously pulling backwards on the core needle.

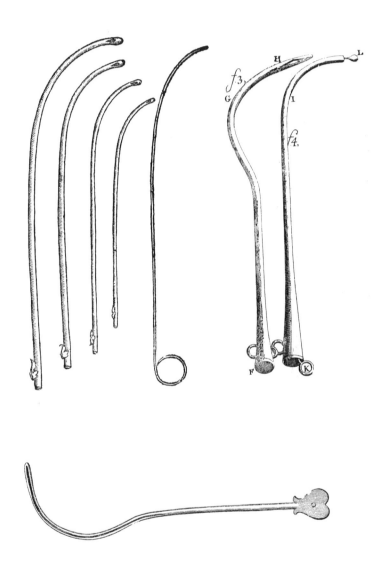

Rigid catheters from 16th-18th century

You obviously realized that the spiral construction had added benefits when you wrote: "The tube is of such a nature that when you have occasion to withdraw [it] its diameter will lessen, whereby it will move more easily...It is also a kind of scrue, and may be both withdrawn and introduced by turning."

We don't know if John ever used the catheter. The next existing correspondence between you two was about a month after the catheter letter. You wrote that John's reply note "gave me a great deal of pleasure, as it informed me that you are better and have reason to think the stone either lessened or made smoother" and mentioned that you prayed to God for a perfect cure.

Clearly, your flexible catheter surpassed the rigid devices then available. By your time, doctors used both straight and curved metal catheters, I've read, for bladder drainage in cases of obstruction. Patients prone to intermittent blockage learned self-catheterization. In some fashionable circles, walking sticks and hatbands contained a small chamber to hold a catheter, ever ready for an emergency.

We don't know if you catheterized yourself; please advise.

Your flexible catheter remains a curiosity in the history of urinary devices, not significantly impacting further developments.

The next big catheter advance came after chemist Charles Goodyear invented *vulcanized rubber* in 1851, a flexible yet tough material created from the sap of a tropical tree; it could easily be molded into a variety of shapes including flexible catheters. (Recall that your friend Joseph Priestley created the word "rubber" because the substance could rub lead pencil marks off paper.)

The last and most beneficial development in bladder catheters occurred in June 1935 when American surgeon Frederick Foley, came up with an ingenious balloon-tipped catheter. His contrivance has two parallel lumens (hollow tubes), one for urine and the other for fluid to blow up a small balloon surrounding the catheter's tip. The Foley catheter completed the evolution of urinary catheters to their present form.

Bladder and kidney surgeons (*urologists*) often use specialized catheters for stone obstruction in a ureter, the passage from kidney to bladder. Some incorporate grasping mechanisms to snag stones blocking outflow.

Today's flexible catheters have an astonishing impact on modern health care. They drain everything from fluid on the brain to infected joints. Urinary catheters also begot heart, arterial and venous catheters.

Certain tests and treatments requiring a flexible device for intervention, especially those demanding a strong twisting action, cannot use a rubber tube because the material isn't stiff enough to withstand torque. In these instances, modern designers came up with the only logical approach to the problem: a spiral strip of metal braised to end caps that prevent uncoiling.

Such devices have shafts nearly identical to your flexible catheter design. They're difficult to use properly because they can kink or untwist. A surgeon, before commencing an operation with a flexible metallic shaft, should read your last words to John in 1752: "Experience is necessary for the right using of all new tools or instruments as well as better direct the manner of using it."

Good advice regardless of the device.

Charles Goodyear

From: "Stuart Green" <stuartgreenmd@yahoo.com>>
To: "Benjamin Franklin" <dr_benjamin_franklin@yahoo.com>
Subject: **A little touch of the gout**

Dear Doctor Franklin:

If you're reading this, you've come to your senses and will want something done about your gout. I know you suffered terribly from the malady. Not long ago I read your little booklet—a bagatelle you called it—printed in 1780 to amuse your friends. Your *Dialogue between the Gout and Mr. Franklin* reveals many truths about gout attacks, but voices 18th-century misconceptions as well. I like best the opening dialogue:

MR. F. Eh! Oh! Eh! What have I done to merit these cruel sufferings?
The Gout. Many things; you have ate and drank too freely, and too much indulged those legs of yours in their indolence.
MR. F. Who is it that accuses me?
The Gout: It is I, even I, the Gout.

We now know that tiny crystals are deposited in joints during a gout attack, causing painful inflammation and swelling. The crystals may also show up in other tissues, but it's the joint pain—commonly the big toe's first joint—that for thousands of years defined gout.

The crystals are a salt of uric acid, a substance discovered by the Swede Karl Scheele (a pioneer of modern chemistry) while you lived in Europe. Although we've no evidence you ever corresponded with Scheele, we know you received a copy of his *Chemical Observations and Experiments on Air and Fire*, sent by a friend.

Usually, when the acute gout attack subsides (after a few weeks to months), a person feels perfectly normal, free of pain, able to go about business and do everything. So it was with you, I believe.

After the first attack, subsequent bouts come on with increasing frequency because uric acid crystals, once deposited in tissues, stay forever. These first crystals, moreover, form a "crystallization nucleus" around which new crystals readily gather, creating wads called tophi; they interfere with joint function.

One fact appears clear to anyone who has studied gout: it affected society's most prominent men—much to the delight of the lowly born! Your own era has been called "The Golden Age of Gout" and you, Dr. Franklin, became one of gout's most illustrious victims. Congratulations.

You obviously knew that high living (meat and alcohol) stimulated gout because of the warning given you by **The Gout** in your own bagatelle. Meat dominated the diet of wealthy Englishmen during the 18th century, perhaps more than any time in history. To make matters worse, British gentlemen washed down their victuals with prodigious quantities of spirits, especially fortified wines like Port and Madeira. As a result, one of every 27 Englishman suffered from gout during your era.

The first mention of your own gout occurred in a 1762 letter to your sister Jane. Do you remember that communication? You told her that you had suffered from "...a little touch of the gout..." but didn't say much else except that "my friends say [gout] is no disease," a commonly held notion at the time.

Physicians prior to your era—and for two thousand years before—considered gout an imbalance of the four humors coursing through our bodies: blood, phlegm, black bile, and yellow bile. That's why physicians hoping to cure gout tried to restore harmony by eliminating the excess humors. This, in turn, led doctors to prescribe bleedings to get rid of surplus blood, diuretics to stimulate urination, emetics to cause vomiting of phlegm, or laxatives and enemas to clear the bowels.

Often laxatives, emetics, diuretics and bloodlettings were combined for maximum effect. (It's a testimony to the human body's resilience that anyone survived a doctor's well-intended care during your lifetime!)

The word "gout" as you know, comes from the Latin *gutta*, meaning, "drop." The ancients assumed a humor dropped into the foot to cause painful swelling. Likewise, since classic gout attacks localized to one joint, physicians prescribed poultices and salves for the affected body part.

Because children seemed immune from the malady, gout ointments sometimes included parts of young animals. (If, Dr. Franklin, you ever used oil of puppy fat or chopped kittens in gosling flesh to ease gout pain, I suggest you keep that information to yourself!)

In 1787, the same year you suffered so miserably from gout while working on the U.S. Constitution, a nephew of your friend William Heberden discovered that tophi are composed of uric acid crystals.

A half century later, Sir Alfred Garrod developed one of the world's first "blood tests" to confirm a high uric acid level. He also recognized the true nature of gout in 1848 when he wrote, "acute gouty arthritis is an inflammatory reaction to crystals of sodium urate deposited in and around the joint."

A diarrhea-producing anti-gout potion recommended by Hippocrates included a flower in the crocus family that caused diarrhea and vomiting. For more than two millennia thereafter, physicians followed the ancient recommendation to administer laxatives and purgatives during an acute gout attack.

In 1780 a medicine appeared in Paris called "L'Eau d'Husson." Did you ever use it? Husson's gout remedy had a secret formula containing powdered Autumn Crocus.

To me, Dr. Franklin, it's inconceivable that you weren't advised to try L'Eau d'Husson for your gout, yet my search of your known documents finds no mention of Monsieur Husson. Perhaps you took a dose or two but didn't like the side effects (nausea, diarrhea, vomiting, buzzing in the ears).

It's also possible that you avoided *all* treatment for gout, thinking the disease was somehow good for you.

The claim that gout prevented more serious ailments became fashionable in England during your era. In some instances, doctors tried to precipitate a gout attack (by plunging a patient's foot in hot or cold water) to cure a deadlier illness like consumption (now *tuberculosis*) or dropsy (we call it *congestive heart failure*) or apoplexy (*stroke*). Your contemporary, the British author Horace Walpole wrote, "I have so good an opinion of the gout...that I do not desire to be cured...I am serious...I believe the gout a remedy and not a disease..."

Did you have the same opinion? I realize that in your *Dialogue* **THE GOUT** said to **MR. F**: "Is it not I who...have saved you from the palsy, dropsy and apoplexy? ...and as to regular physicians, they are at last convinced that the gout...is no disease but a remedy..." Your answer was evasive: "I never feed physician or quack of any kind..."

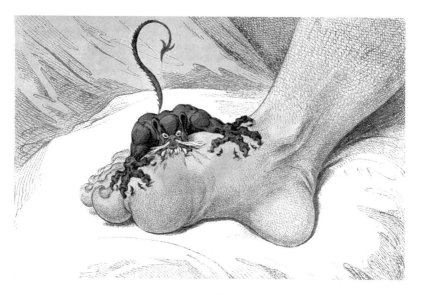

James Gillray

I suspect you doubted the alleged benefits of gout because in 1765 you wrote this tongue-in-cheek letter to your friend John Ross: "If, according to the custom here, I congratulate you on your having a severe fit of the gout, I cannot avoid mixing some condolence with my congratulations: For I too have lately had a visit, or rather a *visitation*, from the same friend (or enemy) that confined me near a fortnight. And notwithstanding the salutary effects people talk of, to comfort us under our pain...I may possibly be, as they tell me, greatly obliged to the gout; but the condition of this obligation, is such that I cannot heartily say, Thank-ye."

35

At the beginning of the 19th century, physicians figured out that the anti-gout benefit of Husson's Elixir came from the bulb portion of the Autumn Crocus, and in 1820, French chemists isolated the active ingredient, which they called "colchicine" from the Latin name of the plant.

Autumn Crocus

From that moment to this, colchicine has remained the most effective method of stopping an acute gout attack. (Several historians believe you brought the Autumn Crocus to America for cultivation for its anti-gout properties.)

Your *Dialogue* still amuses people today. I'm reading it again. I enjoy the part where you tell **THE GOUT**, "…you would not only torment my body to death, but ruin my good name; you reproach me as a glutton and a tipper…"

THE GOUT wasn't easily bullied. He answered, "…If your situation in life is a sedentary one, your amusements, your recreation, at least, should be active…Why, instead of gaining an appetite for breakfast by salutary exercise, you amuse yourself with books, pamphlets, or newspaper, which commonly are not worth the reading."

I laughed out load (LOL in email language) when I read how **THE GOUT** mocked your carriage rides as a "most slight and insignificant" form of exercise, saying that an hour of walking equals in exercise four on horseback and all day in a carriage." I know that you promised **THE GOUT** to "take exercise daily and live temperately" but **THE GOUT** wasn't convinced: "I know you too well. You promise fair; but after a few months of good health, you will return to your old habits; your fine promises will be forgotten like the forms of last summer's clouds."

We all have the same bad habits today, Sir, perhaps even worse than in your times.

From: "Stuart Green" <stuartgreenmd@yahoo.com>
To: "Benjamin Franklin" <dr_benjamin_franklin@yahoo.com>
Subject: **Disabled by the stone**

Dear Doctor Franklin:

At some point after your revival, doctors will recommend removing the bladder stone that troubled you so in your former life. In this email, I'll explain why you should submit to an operation—perhaps one that won't require cutting through your skin.

Because you have gout, your stones are probably composed of uric acid, or at least started out that way. Also, you've a family history of stones. Recall writing nearly forty years before your first gout attack to your "Honoured Father and Mother" with some advice for "stones and gravel." You recommended, "Mrs. Stephens's medicine for the stone and gravel, the secret of which…has for its principal Ingredient, soap…" You also suspected that "honey and mellasses" were helpful because, as heavy liquids, they might, "encrease the specific gravity of our fluids, the urine in particular, and by that means keep separate and suspended therein those particles which when united form gravel."

You later advised your older brother John, "whether it be an ulcer in the passages or a stone I believe onion pottage may be properly taken and to advantage as it lubricates, and at the same time is a dissolvent of calcarious matter."

We know more about *your* stone problems than those of anyone else in the 18th century because of your extensive correspondence on the subject with friends and well-wishers. The detailed letter you wrote to London physicians in 1785 especially impresses historians. Your use of the third person, as I've said before, seems quaint to modern ears.

Do you remember what you told the doctors? "When a young man he was sometimes troubled with gravelly complaints; but they wore off without the use of any medicine, and he remained more than fifty years free from them." In the summer of 1782, when seventy-six, according to the letter, you experienced "a severe Attack accompanied with what was thought to be a gouty pain in the hip…" and that you, "daily voided gravel stones the size of small peas" for which you took honey, about a pound's worth in a week.

After that attack you started "observing sand constantly" in your urine. About a year later, you felt pain while traveling in a carriage and began to have problems voiding: "At times when he was making water in full stream, something came and stopt the passage; this he suspected to be a small stone, and he suffered pain by the stoppage. He found however by experience that he could by laying down on his side cause the obstruction to remove and continue the operation."

Considering the dangers of surgical treatment for bladder stones in your era, most people would agree that you rightly "would choose to bear with it rather than have recourse to dangerous or nauseous remedies." You obviously sought two answers from the British physicians: 1) how to keep the stone from enlarging; and 2) how to ease the discomfort.

18th-century
lithotomy
knife

Being a doctor myself, I understand why your London consultants concluded "the symptoms are not so urgent as to render life uncomfortable, we approve much of his resolution not to risk an operation, or any course of medicine that might endanger his health."

Having said that, I can also understand your frustration when after a life of vigorous activity, the physicians recommended that you stop exercising and confine yourself to your house and garden, and to "avoid all riding in a carriage."

Lithotomy (surgical stone removal) is one of the oldest surgeries in existence. With the patient in the "lithotomy position" (still used today for pelvic examinations of women) the lithotomist made an incision into the bladder and extracted the stone with a hook. If bleeding could be stopped by compression and the subsequent infection remained localized, a patient might survive the ordeal

You displayed a stoic outlook in a letter to your protégée, the rabble-rousing pamphleteer Thomas Paine: "I have the stone indeed and sometimes the gout, but the pain from the stone is hitherto not very severe; and there are in the world so many worse maladies to which human nature is subject, that I ought to be content with the moderate share allotted me."

Obviously your stone symptoms worsened because the following year (1786) you decided against purchasing a country plantation, "since my malady, the stone, does not permit me to ride either on horseback or in a wheel carriage."

You no doubt tried medicines to dissolve your stone, Sir, because you seemed particularly discouraged by the failure of numerous oral remedies. The letter you wrote to Scottish Army surgeon Alexander Small fourteen months before your encaskment contained a particularly prophetic statement about bladder and kidney stones: "I have no hope of it being dissoluble by any medicine."

Sadly, Dr. Franklin, no effective dissolvent exists, either then or now. So you see, not everything has progressed.

Considering the intense bladder pain you endured, Americans applaud your fortitude for attending every session of the 1787 Constitutional Convention, even though lugged there in your European carrying chair by sturdy prisoners.

I only wish more people today faced the infirmities of old age with the same equanimity you displayed when you wrote a friend, "people who live long, who will drink of the cup of life to the very bottom, must expect to meet with some of the usual dregs; and when I reflect on the number of terrible maladies human nature is subject to, I think myself favoured in having to my share only the stone and gout."

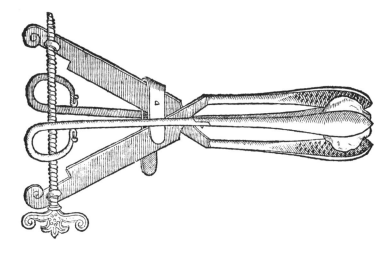

Stone removal instrument from 18th century

In recent years, doctors have tried different ways to break up bladder stones. For example, Russian surgeons invented a contraption that splits stones with electrically generated pressure waves inside the bladder. Some modern fragmentation strategies use sound waves or even strong light beams (called *lasers*) to break up small gravel in the ureters. If a modern surgeon finds a large bladder stone, he or she will simply remove it surgically.

Fear not, Dr. Franklin, modern operations are painless. Do you recall the *ether* evaporation experiments you performed with Professor John Hadley at Cambridge in the spring of 1758? That substance has a remarkable property: inhaling its vapors induces a deep narcosis—greater even than tincture of opium—with rapid emergence when the fumes are removed.

United States Patent Office.

C. T. JACKSON AND WM. T. G. MORTON, OF BOSTON, MASSACHUSETTS; SAID C. T. JACKSON ASSIGNOR TO WM. T. G. MORTON.

IMPROVEMENT IN SURGICAL OPERATIONS.

Specification forming part of Letters Patent No. 4,848, dated November 12, 1846.

Patent application for ether use during surgery

In the 1840s, two American doctors began using ether to create a state of *anesthesia* prior to starting their operations, thereby ushering in the modern era of surgery. So don't despair any longer about your bladder stone; surgeons will have it in a bottle shortly—and you'll be relieved.

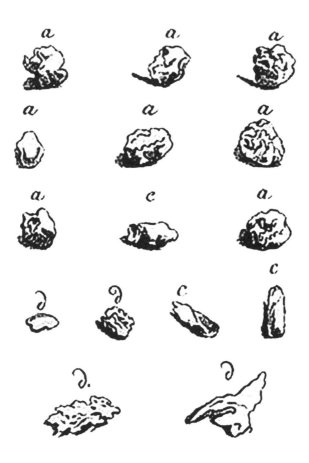

Kidney and bladder stones

Measures developed in the mid 19th century have greatly reduced post-operative complications. First in Vienna and later in London, insightful physicians realized that their own dirty hands and instruments spread infections from one patient to another. Hand washing with soap—the same kind your father made in his tallow shop—prevents such transmittal.

In a future email, I'll tell you about how *antibiotics*—one of the great discoveries of the 20th century—also help prevent operative infections.

From: "Stuart Green" <stuartgreenmd@yahoo.com>
To: "Benjamin Franklin" <dr_benjamin_franklin@yahoo.com>
Subject: **Scarcely escape newspaper defamation**

Dear Doctor Franklin:

Because you're the first revitalized person, few will believe announcements of your unearthing and revival. People will want proof of such an astounding occurrence. Skeptics will claim that Philadelphia's officials faked the story of your discovery for publicity purposes. Members of the press, by the time you read this, will have established a vigil outside your hospital window, awaiting a glimpse, a word, anything tangible to share with the world.

At some point after you reawaken, Sir, you must meet with the press to prove your identity. This won't be easy. Simply declaring, "I am Dr. Benjamin Franklin of Philadelphia" will not do. Instead, you must supply details about your life and your family and your accomplishments to convince even the most cynical disbeliever of your return. We're all naturally doubting Thomases.

I recommend, however, that you wait a while before granting any interviews—to regain your strength, reacquaint yourself with the history of your times, and become familiar with the modern vernacular. My emails will, I hope, help.

Once you feel ready, you have two ways to make yourself known to the public. One, a *press conference*, is a large gathering of reporters shouting out their questions without orderliness or courtesy. The experience will remind you of the unpleasant time you testified in the House of Commons to get the Stamp Act repealed. (Although you had "friends" among the Members of Parliament who asked "planted" questions, recall as well the "enemies" who tried to humiliate you with challenging queries.)

The other format available to convince the public of your identity is the *personal interview*. In this situation, one person asks all the questions, which you answer. The public watches over the interviewer's shoulder, so to speak, on a device similar to the one in front of you.

You must, Sir, be wary of the press. Their desire to gain fame by attacking others hasn't changed from your time to ours. Recall you once predicted that even if the Angel Gabriel reappeared, "he could scarcely escape newspaper defamation from a gang of hungry ever-restless, discontented and malicious scribblers."

I recommend that you avoid the press conference and select the personal interview instead. The commotion during a press conference will prove so distressing you'll wish you were back in your barrel. Also, you're reportedly uneasy when speaking in public but have a favorable reputation as conversationalist in private discourse.

Madame Helvetius

Although many are qualified to conduct a meaningful interview, I suggest that you consider one of several ladies who have become famous for their interrogatory skills. One, with an agreeable nature, will remind you of your Paris friend Madame Brillon, while the second being closer to your age, will take you back to Madame Helvetius. The first has a warm interview technique that will prove quite comfortable, whereas the second is more abrupt. Indeed, her first question (after exchanging pleasantries) might be, "Tell me Benjamin, who was the mother of your illegitimate son William?"

Nevertheless, you might prefer the second because, being divorced, she is, shall I say, available. An interview with her could turn into an interesting match-up, like the chess games you so enjoyed during your lifetime: she would battle to corner your King, while you campaign to topple her Queen.

From: "Stuart Green" <stuartgreenmd@yahoo.com>
To: "Benjamin Franklin" <dr_benjamin_franklin@yahoo.com>
Subject: **Hide not your talents**

Dear Doctor Franklin:

Even though you've earned no income for 225 years, you needn't worry about living expenses. Although your Will distributed your considerable fortune, as a now-living person you have the right to compensation for the use of your name and likeness.

Since your proclaimed death, numerous banks and financial businesses named *Franklin* have appeared, many capitalizing on your reputation for thriftiness and organization. Moreover, every one of the fifty states has either a county or town named after you, and because most cities have a Franklin Street, Road or Avenue, countless small businesses also carry your name. Some companies might claim that "Franklin" doesn't refer to you, but many Franklin businesses picture your face on their documents and advertisements, so they must either change their name or pay you royalties.

I can recommend a lawyer to negotiate such payments, which I assure you will be substantial.

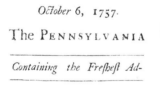

October 6, 1757.

The PENNSYLVANIA

Containing the Freſheſt Ad-

NUMB. 1502:

G A Z E T T E

vices, Foreign and Domeſtick.

Masthead from your newspaper

You should also consider completing the written story of your life, the one you never finished while alive. The version you penned for son William ended in your 44th year, even though you planned to conclude it before you died. The part you did complete has sold briskly as *The Autobiography of Benjamin Franklin* for more than two centuries, in many dozens of editions.

Since so many important things happened during your life's next 40 years, an *Autobiography Part II* would, I predict, become a *best seller*. Publishers will offer you an *advance* for the book—a large amount of money in anticipation of sales. Likewise, individual chapters can be sold to periodicals.

Before putting pen to paper, however, I suggest you read *all* my emails along with a few general biographies of your life, to refresh your memory of things past.

As a retired political person, you'll have many offers for financial remuneration. For instance, you might join a *Speaker's Bureau*, giving lectures to groups looking for a famous person to enlighten their members. For someone of your renown, the fee paid for one forty-minute speech would exceed the annual salary of a judge or chief of police. (I suspect you'll easily overcome your reluctance to speak in public for that kind of money!) Since the sponsoring group pays the speaker's travel and lodging expenses, you'll visit many parts of the country—even the world—in exchange for a few words from *Poor Richard's Almanack*.

It must appear obvious you'll have no difficulty making a substantial income provided you follow *Poor Richard* advice: "Hide not your talents, they for use were made; What's a sun-dial in the shade?"

My next group of emails will cover a subject of considerable interest to you—microscopy. I will focus (pun intended) on the work of Robert Hooke, whom you so admired.

Robert Hooke

44

MICROSCOPY & CONTAGEOUS DISEASES

That admirable instrument the MICROSCOPE 47

Air is replete with infinite multitudes of living creatures 49

An account of the small-pox 51

Suspected to propagate infectious distempers 57

The same distemper is often bred in ships 61

Take one ounce of good Peruvian bark 63

The yellow or bilious fever 67

A continual risque to my health 69

Great numbers of you catch colds 71

A disgrace among the ancient Persians to cough or spit 75

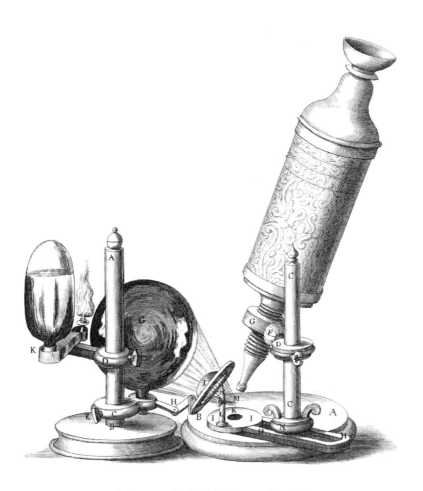

Microscope *from* Hooke's Micrographia, 1665

From: "Stuart Green" <stuartgreenmd@yahoo.com >
To: "Benjamin Franklin" <dr_benjamin_franklin@yahoo.com>
Subject: **That admirable instrument the MICROSCOPE**

Dear Doctor Franklin:

In this email I'll tell you about advances in microscopy, a subject that obviously captured your attention.

Do you remember back to 1751 when your *Poor Richard's Almanack* contained an account of "That admirable Instrument the MICROSCOPE"? You enlightened readers about invisible wonders discovered by Anton Van Leeuwenhoek—father of microscopy—and British scientist Robert Hooke. Consistent with the style of your era, you credited neither of them when *Poor Richard* wrote, "The globules of the blood, which are computed to be almost a two thousandth part of an inch in diameter" and bones, when sliced thin consist of "innumerable perforations, and ramifications, disposed in an endless variety of forms."

Human skin, you said, "is found to be covered over with an infinite number of scales lying over one another, as in fishes; and it is probably the same in other animals."

You erred, however, when you said, "a great drone fly [has] no less than fourteen thousand Eyes, with a distinct optic nerve to each; and each eye appears through the microscope, tho' magnified many hundred thousand times …" The instrument's magnification factor was, in truth, but a few hundred-fold.

I was pleased to see your section about the polyp—that tiny creature whose "young ones come out of the sides of the old, like buds and branches from trees, and at length drop off perfect polypes." You pointed out that the polyp "when cut into a great many pieces, each several piece becomes a complete polype."

"A bit of cork cut extremely thin," said *Poor Richard*, "in the microscope, are so many curious pieces of mosaic work." (Hooke called individual units of tissue *cells*, a word we still use.)

"Moldiness upon flesh, leather, or other substances, is no other than a great number of extremely small, but perfect plants, having stalks and tops like mushrooms, and sometimes an appearance of leaves," wrote *Poor Richard*, "The seeds of these minute plants must, in all probability, be diffused universally through the air, and falling upon substances fit for their growth, spring up in astonishing profusion."

It's paragraph number seven, however, that intrigues the most: "By the help of the microscope the innumerable and inconceivably minute animalcules in various fluids are discovered, of the existence of which we have no reason to suppose any mortal had the least suspicion, till last century." *Poor*

Richard proclaimed, "Of a certain species some are discovered so extremely minute, that it has been computed, three millions of them, or three times the number of the inhabitants of London and Westminster, would not equal the bulk of a grain of sand."

You next characterized the micro-organisms' appearance: "Of animalcules, some species resemble tadpoles, serpents or eels, others are of a roundish or oval form, others of very curiously turned and various shapes; but in general they are extremely vigorous and lively, and almost constantly in motion." You here described *bacteria* in both spiral and globular forms, adding, "Animalcules are to be found...in the foulness upon our teeth, and those of other animals, in our skins when affected with certain diseases..."

Did you suspect that infestation by microscopic organisms actually *caused* foulness or disease? Likewise, did you understand the destructive potential of the tiny animalcules—as we say, *germs* or *microbes*—when you informed readers that "These vermin sometimes...will eat up the head and part of the body of a polype, after which, if it be cleared of them, it shall have the devoured parts grow up again, and become as complete as ever"?

For 180 years after your *Almanack* article, microscopy slowly improved without a major technological advance. Progress came as a result of competition-driven manufacturing improvements rather than by theoretical insights.

Improved optics, however, couldn't enhance magnification more than four-fold beyond what you described in 1751. Here's why: the maximum possible magnification with a perfect optical microscope is 1250X, limited by the nature of light itself.

To probe deeper into the invisible, scientists in 1931 discovered how to focus with magnets infinitesimally small particles of the single "electric fluid" (now *electrons*) that you characterized in 1750.

The modern *electron microscope* magnifies to a million fold. For this reason, it strikes me as a remarkable coincidence that in the same month *Poor Richard's* readers contemplated the MICROSCOPE, your London friends published your book *Experiments and Observations on Electricity made at Philadelphia in America*.

And the rest, as we say, is history.

Mold from Hooke's *Micrographia*

From: "Stuart Green" <stuartgreenmd@yahoo.com >
To: "Benjamin Franklin" <dr_benjamin_franklin@yahoo.com>
Subject: **Air is replete with infinite multitudes of living creatures**

Dear Doctor Franklin:

Do you remember corresponding with your British friend Dr. John Pringle (in 1777, shortly after arriving in Paris) about your boils? You wrote in the third person—as though speaking of someone else—and didn't mention that you were at the Continental Congress when your rash first appeared.

You told Sir J.P. that "Towards the end of winter 1776, he set out on a journey of 500 Miles, of which great part was performed in a small open boat, where he was kept sitting without exercise for many days."

Do you recollect that failed effort soliciting Canadians to join America's rebellion? While heading up the Hudson River and across Lake Champlain, you said you were "afflicted with a succession of boils, sometimes two or three together, each when healed leaving round about it spots of the same scurff, which obstinately continued, being renewed after every removal."

Fig. 1 Scurf   *Fig. 2 Scales*   *Fig. 3 Crust*   *Fig. 4 Scabs*

Skin lesions from an 18th-century medical textbook

Your next assignment sent you to France to raise money for the Revolutionary War. Again, you said nothing to J.P. about the trip's purpose: "In November 1776 he made a long sea voyage, in which the disorder sensibly increased, and the boils became more frequent. Part of each arm, and of each side, the small of his back, and parts of his thighs and legs, became covered with the scurff, which became very troublesome, itching sometimes extremely, and when rubbed or scratched off, would spot his linen with blood."

We now know that boils, empyema, and suppuration are different manifestations of the same thing, all caused by the "kind of vermin" you mentioned in your MICROSCOPE article.

For millennia before your time, *spontaneous generation* explained the appearance of numerous kinds of creatures in rotting trees, muddy swamps, and all manner of decay. Overthrowing that doctrine proved essential to establish a relationship between the microscopic vermin and disease.

In 1668, the Italian scientist Francesco Redi (whose book you owned), placed two identical pieces of meat in separate jars, leaving one open and while covering the other with a snug-fitting cloth. Maggots appeared only in the uncovered jar.

You obviously knew of Redi's work. Recall telling *Almanack*'s readers that no microscopic animals appear in fluids if they remain covered, "but when open to the access of the air, their numbers are beyond reckoning..." You even explained how this happened: "Air is replete with infinite multitudes of living creatures too small for sight...which fall upon places fitted for them..."

For the next hundred years after your MICROSCOPE piece, occasional publications described microorganisms in diseased tissues. Today, however, we credit the experimental work of French chemist Louis Pasteur and of German physician Robert Koch for our modern understanding of infectious diseases.

Pasteur first studied wine spoilage. He suspected that microscopic animalcules caused the problem, so he conducted experiments that buried forever the notion of spontaneous generation. He proved that the tiny organisms arrived at the wine's surface by settling out of the air.

Pasteur quickly realized that he could prevent spoilage by keeping airborne microbes away from wine and other beverages after he first heat-killed bacteria already there. *Pasteurization* with heat has since become mandatory in the beverage industry.

Asked to investigate anthrax (splenic fever of sheep and cattle), Pasteur determined that a microbe caused the disease, but more significantly, found he could stimulate permanent immunity by inoculating sheep with *killed* anthrax bacteria. He soon applied the same principle to rabies, even though he never found a bacterium that caused the illness. Without knowing about existence of even smaller microscopic beings called *viruses*, Pasteur heated an extract from a rabid dog's brain to inactivate the infectious organisms, creating an inoculation that prevented rabies.

Robert Koch also started his scientific career studying anthrax: he advanced the techniques used today to isolate and identify bacteria; he discovered the tuberculosis bacillus (1882) and the bacterial cause of cholera (1885). He made contributions to our present understanding of malaria, typhus, and other diseases I'll tell you about in a couple of days. Most significantly, Koch established the idea that each microbiological species causes its own distinct disease.

Based on your selection of material for the *Almanack*, it appears that you accepted—or at least considered—the notion that microscopic creatures could cause infectious contagions in persons so predisposed. Today, thanks to countless researchers, we suffer less from invasion by "minute animalcules" than from our own excesses—at the dinner table, on the barstool, in the armchair.

Understanding the cause of an illness and achieving a cure are, of course, two different things. With respect to diseases produced by bacteria, a fortuitous accident in a British laboratory led to the discovery of *antibiotics*—medicines that kill microbes without harming the patient. I'll tell you more about them after I review some "infectious distempers" from your era.

From: "Stuart Green" <stuartgreenmd@yahoo.com>
To: "Benjamin Franklin" <dr_benjamin_franklin@yahoo.com>
Subject: **An account of the small-pox**

Dear Doctor Franklin:

Although you may find it impossible to believe, smallpox no longer exists. Of all the afflictions that plagued mankind, none affected you as deeply as the smallpox epidemic of 1736, which carried away your 4-year-old son Francis (Franky).

Your 1772 letter to your sister Jane grieves me when you spoke of "my son Franky, tho' now dead 36 years, whom I have seldom since seen equaled in every thing, and whom to this day I cannot think of without a sigh."

Rumors swirled through Philadelphia, you'll recall, that Franky died from a smallpox inoculation. You tried to set the record straight by publishing an announcement in your *Pennsylvania Gazette* declaring that Franky "was not inoculated but received the distemper in the common way of infection."

I also know you truly intended to have Francis inoculated, but delayed because he had another illness. Having put off Franky's inoculation must have added a double sense of loss when he died, first because you no longer had your son, and second, because the death might have been avoided if you only followed your "known opinion that inoculation was a safe and beneficial practice."

I can understand a parent's reluctance to inoculate by the method used in your times. After all, as performed during the 18th century, the patient was purposefully given a mild case of smallpox to immunize against a more serious one in the future.

A highly contagious virus causes smallpox. (Viruses are the tiny organisms—hundreds of times smaller than the bacteria I mentioned in yesterday's email.) About thirty percent of those who contract the disease perish during the illness as the virus invades their internal organs. The smallpox virus, like all other viruses, cannot multiply on its own; instead, it takes over the reproductive machinery of the invaded cells to create copy viruses that soon attack other cells. The rapidly multiplying viruses eventually overwhelm the victim.

Virus particles, when suspended in the saliva of the infected, pass to others when victims cough or sneeze. Contact with the fluid from infected skin blisters also spreads the disease.

The disease got its name from the small pock-marks (skin pits) that cover victims who survive the illness. Once infected, a survivor gains lifetime immunity from smallpox. This well-known fact suggested that a pock-marked slave offered for sale would be a wise purchase. Do you remember that your

Pennsylvania Gazette, in 1732, carried an advertisement for a "young negro fellow, about 19 or 20 years of age" who was "very fit for labour, being used to plantation work, and he has had the small-pox"?

Many years ago, the Chinese discovered that one could acquire a mild form of smallpox—thereby providing lifetime immunity—by inhaling ground-up scabs taken from sufferers of the disease. The resulting illness was usually mild, although about 3% of those so infected didn't survive the ordeal.

(I realize you knew about the Chinese inoculation for smallpox. In 1746 you told your Boston friend William Vassall that you had "somewhere read that the Chinese actually preserve scabs taken from a healthy person for the purpose, tho' their manner of inoculation is different from ours.")

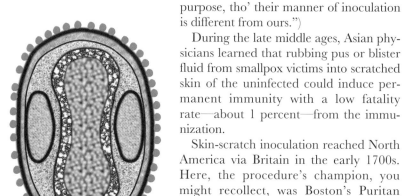

Smallpox virus, cross-section X100,000

During the late middle ages, Asian physicians learned that rubbing pus or blister fluid from smallpox victims into scratched skin of the uninfected could induce permanent immunity with a low fatality rate—about 1 percent—from the immunization.

Skin-scratch inoculation reached North America via Britain in the early 1700s. Here, the procedure's champion, you might recollect, was Boston's Puritan leader Cotton Mather. There were also local voices opposed to inoculation—people whom though it absurd to induce a potentially fatal illness to prevent one. Recall that a particularly vocal member of the opposition was your own brother James, publisher of the *New England Courant* and employer—technically the owner—of an indentured apprentice, namely you.

Do you remember the *Courant's* first issue of August 7, 1721? It contained an attack on smallpox inoculation written by Dr. William Douglass, the only physician in North America with a European M.D. degree. His article heaped "Ridicule on the inoculator" for "the dubious, dangerous practice of inoculation."

The Mathers struck back with a factious *Anti-Courant*—obligingly printed by your brother James and perhaps typeset by you—which ranted, "Go on, Monsieur Courant, and prosper; Fear not to please your stupid admirers...and write in your native stile...VERY, VERY DULL!"

The *New England Courant's* next issue contained a further assault, setting the stage for a running battle between your brother and the Mather family. Cotton Mather called the debate "a libel on purpose...despicable, even detestable...a wickedness that was never known before in any country, Christian, Turkish or Pagan, on the face of the earth." After that, your brother backed off and turned his paper towards lighter fare.

According to your *Autobiography*, you gradually tired of James' increasingly "harsh & tyrannical" mistreatment. Your sly escape from Boston, ending in Philadelphia at the age of 17, has become an American legend.

Nobody today knows your thoughts about the smallpox inoculation controversy that permeated Boston while you lived there. Ten years later, however, you came down squarely on the side of inoculation. Your *Pennsylvania Gazette* in 1730 described a Massachusetts epidemic thus: "There is an account published of the number of persons inoculated in Boston in the month of March, amounting to seventy-two; of which two only died, and the rest have recovered perfect health. Of those who had it in the common way, [that is, by what we now call person-to-person spread] 'tis computed that one in four died." Likewise, your letters to sister Jane often containing news of smallpox epidemics, reminding her to get inoculated.

Twenty-three years after arriving in Philadelphia, you had acquired enough information about the safety of inoculation to tell Vassall that "between 150 and 160 persons... have been inoculated... of which number one only died."

I know that you advocated self-inoculation for the poor or for those without suitable access to a doctor because you told Vassall how to do it: "As to your going to New York to be inoculated, perhaps such a journey is not quite necessary; since, as has been tried here with success, a dry scab or two will communicate the distemper by inoculation, as well as fresh Matter taken from a Pustule and kept warm till applied to the incision. And such might be sent you per post from hence, corked up tight in a small phial." Nobody knows if you were thus inoculated—perhaps you'll enlighten us when you get a chance.

Although you couldn't have known it at the time, while you drafted your proposal to create the Philadelphia Academy (now the *University of Pennsylvania*), a boy named Edward, destined to conquer smallpox, was born to Vicar and Mrs. Stephen Jenner of Berkley, in the borough of Gloucestershire, about 150 miles west of London.

At the age of seven, Edward Jenner nearly died of a bad reaction to a smallpox inoculation. He never forgot the experience. Six years later, when only thirteen, Jenner was apprenticed for seven years to a physician, your London friend Dr. John Hunter. Jenner noted that clear-skinned milkmaids with cowpox pimples on their hands seemed strangely immune from smallpox even though they had never been inoculated. Years passed before Jenner grasped the importance of this observation.

Meanwhile, debates raged about inoculation with fluid from smallpox skin lesions. In our own time, when parents worry about exposing their children to a possibly of a fatal reaction that occurs once in 10 million immunizations, it's easy to understand resistance to inoculation with a 1 in 50 chance of dying from the procedure.

In 1759, the pamphlet you and Dr. William Heberden published in London, *Some Account of the Success of Inoculation for the Small-Pox in England and America,* did more to promote North American acceptance of inoculation than any other public effort.

(Physicians today recognize Heberden's name because it's associated with the small bumps—Heberden's nodes—that frequently develop on the top of our finger joints as we age. He also first characterized angina pectoris—chest pain associated with insufficient blood supply to the heart muscle.)

Your *Preface* to Heberden's smallpox pamphlet impresses modern physicians because it contained statistical evidence that inoculation was much safer than risking infection in "the common way." (Indeed, historians believe that you persuaded Heberden to write the smallpox account to convince reluctant North American colonists to acquire smallpox immunity from inoculation.)

During the next three decades you had little, if anything, to say about smallpox, busy as you were with other matters. At the time of your "death" in April 1790, you couldn't have known that the potential for smallpox eradication was literally at hand.

After Edward Jenner's apprenticeship, he had established a successful if not particularly prominent medical practice. All that changed on May 14, 1796, when Jenner used material from a cowpox lesion on a milkmaid's palm to inoculate an eight year-old boy who never had smallpox. The lad developed a small inoculation-site blister that healed completely in two weeks. After that, Jenner couldn't induce any sort of reaction in the child with blister fluid from a smallpox patient.

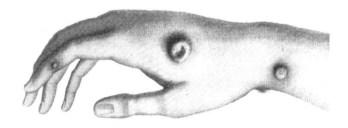

Cowpox lesions by Jenner

Edward Jenner worked tirelessly to promote cowpox inoculations. He called his technique "vaccination" because he used material obtained from skin lesions on cows (*vaccus* in Latin). Jenner had so much confidence in the value of his discovery that in 1801 he proposed a worldwide vaccination program to eliminate smallpox forever.

Thomas Jefferson, when president of the United States, wrote Jenner in 1806, offering "a portion of the tribute of gratitude due to you from the whole human family. Medicine has never before produced any single improvement of such utility...You have erased from the calendar of human afflictions one of its greatest..."

The success of smallpox vaccination seems especially remarkable because the procedure was practiced without understanding the disease's cause. Physicians in your time and Jenner's surmised that smallpox was spread by a toxin—a poisonous substance that, upon entering the body, induced the formation of more toxin, which spread the illness to others. We now know that the toxic material consists of microorganisms that self-replicate at the expense of cells they invade.

Edward Jenner

Today, with the safety of inoculation for many diseases well established, a few parents still fear exposing their children to any risk of an unfavorable inoculation reaction no matter how slight. For this reason alone, they should read your Preface to Heberden's pamphlet: "As the practice of Inoculation always divided people into parties, some contending warmly for it, and others as strongly against it…" it was necessary to have "a strict and impartial enquiry" into the inoculation and death rates during epidemics.

To refresh your memory, here's what you said about the risk of inoculation: "if the chance were only as *two* to *one* in favour of the practice among children, would it not be sufficient to induce a tender parent to lay hold of the advantage? But when it's so much greater, as it appears to be by these accounts (in some even as *thirty* to *one*) surely parents will no longer refuse to accept and thankfully use a discovery GOD in his mercy has been pleased to bless mankind with..."

William Heberden

A worldwide effort to eradicate smallpox started in earnest in 1958 with a comprehensive global vaccination program. On December 9, 1979, the goal was finally achieved: Smallpox is no more. Your own effort to encourage acceptance of inoculation contributed to that objective; never again should a parent lose a child, as did you, to the dreaded scourge of smallpox.

From: "Stuart Green" <stuartgreenmd@yahoo.com >
To: "Benjamin Franklin" <dr_benjamin_franklin@yahoo.com>
Subject: **Suspected to propagate infectious distempers**

Dear Doctor Franklin:

Do you remember meeting on February 24, 1748 with fellow members of Philadelphia's Common Council? On the agenda was "the swamp between Budd's buildings and Society Hill," which you viewed as a "nuisance …filling our own inhabitants with fears and perpetual apprehensions, while it is suspected to propagate infectious distempers…"

I've read that most people during your time thought that swamps emitted a harmful substance that caused contagious diseases. Likewise, they believed that decaying matter and foul-smelling things spread vapors into the air—especially at dusk—that made people ill, particularly in certain seasons of the year. I've seen the word *miasma* used to describe these vapors.

My Bailey's 1757 Dictionary defined miasma as "a contagious infection in the blood and spirits, as in the plague etc. more particularly such particles or atoms as are supposed to arise from distempered, putrifying or poisonous bodies, and to affect people at a distance."

In your era, people feared foul air, especially when near cemeteries and swamps. Many believed they could see miasmic vapors, as either a "dense blue mist" at dusk or an "evil, malevolent, contagious, destructive" yellow fog in the mornings. People, as you know, stayed indoors during epidemics; they slept with windows closed, preferring stuffy indoor air to the poisonous haze beyond their walls.

If I correctly understand your position, Sir, you preferred inhaling fresh air from outdoors rather than continuously rebreathing indoor air, clearly a minority viewpoint at the time.

A famous anecdote about your insistence on sleeping with the windows open comes from your onetime bedfellow, John Adams. Recall that Congress, after declaring independence in Philadelphia, sent you and Adams north to discuss a possible truce with British General Howe before hostilities with the mother country worsened.

A waypoint New Jersey inn proved so crowded, according to Adams, you both shared a bed in a tiny room with one window. As Adams tells us, "The window was open and I who was an invalid and afraid of the air of the night, shut it close. 'Oh!' says Franklin, 'don't shut the window; we shall be suffocated.' I answered I was afraid of the evening air. Dr. Franklin replied: 'The air within the chamber will soon be, and indeed is now, worse than that without doors. Come, open the window and come to bed, and I will convince you."

Adams said you so regaled him with your theory about catching colds that he soon fell fast asleep.

The idea that poisonous miasmas caused contagious diseases, while seemingly foolish in light of modern microscopy, served an important public health function. It compelled officials to eradicate sources of miasmic mists—the swamps—although for the wrong reasons. (We now know that swamps breed disease-transmitting mosquitoes.)

To eliminate Philadelphia's Society Hill swamp, you and your fellow Council members were wise to contract with adjacent property owners to dredge "the dangerous Nuisance" down to the river—their expenses reimbursed by future docking fees.

As the discoveries of Pasteur, Koch and other experimenters in the mid-1800s gained acceptance, physicians everywhere began to look for—and find—microscopic organisms associated with infectious and contagious illnesses.

Since I need precise terminology, Dr. Franklin, consider this: An *infectious disease* occurs when organisms invade and harm our tissues—the invader may be as small as a virus or as large as a twenty-foot tapeworm—but not all infectious diseases are contagious. A farmer who develops gangrene after puncturing his foot with a dirty horseshoe nail will not transfer that infection to those caring for him. Boils originate with *microbes* (another term for microscopic organisms) already on our skin; we don't catch them from others.

Contagious diseases are infections spread from person to person either by close contact or by intermediate carriers. Drinking water, food, insects, and household objects contaminated by those already infected can transport germs from the sick to the healthy.

Smallpox, as I said previously, transmits by contact or by coughing. The only way officials of your era could control such contagious illnesses that spread from person to person—quarantining the infected—often proves difficult and costly. Alternatively, officials prohibited public gatherings during epidemics. For example, remember this announcement in your April 22, 1731 *Pennsylvania Gazette*: The Justices of the Peace in nearby Burlington "by reason of the great mortality in Philadelphia, and other parts of Pennsylvania, where the small-pox now violently rages...to prevent...the further spreading of so epidemical and dangerous a distemper" prohibited the annual May Fair in their town.

Cholera spreads via water contaminated with the feces of individuals already ill with the disease. A bacterium, *Vibrio cholera*, causes the often-fatal illness by infecting the bowels, inducing severe dysentery characterized by vomiting and especially, diarrhea.

Cholera epidemics became increasingly common during your century and the following one when cities became overcrowded yet didn't have proper human waste disposal systems. As a result, in London, New York, and other teeming cities, cholera epidemics swept through certain neighborhoods every ten years or so.

A London physician, Dr. John Snow, analyzed cholera cases during an 1854 epidemic, tracing the disease to a single well on London's Broad Street.

John Snow

Snow was convinced that a germ contaminated the Broad Street well, and found Vibrio cholera bacteria in the water.

Snow ended the epidemic by removing the pump handle, a measure that earned him a knighthood—and lasting fame as the father of *epidemiology* (the study of epidemics).

We now know that certain distempers so feared in colonial Philadelphia—typhus, bubonic plague, yellow fever, and malaria —are among many contagious diseases transmitted to humans by insects. I'll take you back you to these illnesses in my next three emails.

John Snow's map of London's 1854 Cholera epidemic

From: "Stuart Green" <stuartgreenmd@yahoo.com>
To: "Benjamin Franklin" <dr_benjamin_franklin@yahoo.com>
Subject: **The same distemper is often bred in ships**

Dear Doctor Franklin:

I realize that you, as civic leader, worried about the **typhus epidemics** that often entered Philadelphia on ships. Like others of your time, you had no idea that *lice* (more spiders than insects) transmit the causative germs. Scratching the louse bite introduces the microbes into the skin. The typhus germ is neither bacteria nor virus, but something in between, a Rickettsia, named for their discoverer, Dr. Howard Ricketts.

For ten days after a person rubs the rickettseal germs into louse bits, he or she feels fine while the microbes multiply. The patient then becomes ill with high fevers and chills, followed by a generalized sick feeling with headaches. A rash the covers the body, and when fatal, the patient becomes delirious and then comatose. Of those so infected, about half die.

A louse cannot fly or jump, so it transfers from person to person by close contact. Body lice proliferate on unwashed skin by laying eggs called nits, which stick to hair and clothing fibers. In yours and the following century, typhus commonly occurred on crowded ships (*Ship fever*), in jails (*Gaol fever*), and in military camps (*Camp fever*).

You helped control the disease by reprinting (in your September 4, 1755, *Pennsylvania Gazette*) a description of Jail Fever written by Dr. John Pringle. I was particularly impressed with the short introduction you wrote for the article, especially when you mentioned "how infectious that distemper is, which in England is called the GAOL FEVER; as it arises there from a number of persons confined together in close gaols, till they have poisoned the air they breathe, and one another…The same distemper is often bred in ships crowded with passengers."

An epidemic would likely spread, you rightly claimed, in "gaols on board the ships, by the felons sent over to America" as indentured servants. Such servants, you pronounced, were procured by "scumming the gaols in Germany." Alternatively, some Philadelphians died by "purchasing Dutch servants out of sickly ships, and bringing them into their houses." You offered wise public health advice when you recommended that your readers be "more cautious how they buy the [typhus] plague, and bring it home to their families."

Bubonic plague ordinarily infects rats; their fleas transmit plague bacteria to humans. As with smallpox, once the germs enter the lungs and salivary glands, coughing spreads the infection from one person to another.

Buboes are painfully swollen lymph nodes, our body's natural defense to the infection's spread; they appear first in the groin because most flea bites are on the legs. Most often, however, lymph nodes fail to stop invasion to deeper organs, resulting in high fevers, confusion, coma and death.

Modern medicines are especially effective against plague germs. Likewise, control of rat populations has eradicated plague transmission in most places.

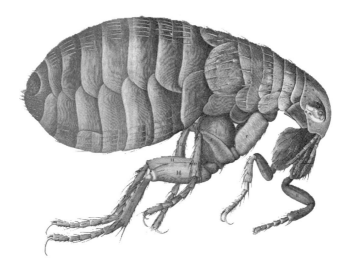

Flea from Hooke's *Micrographia*

Fortunately for you and your family, America during your time didn't experience bubonic plague epidemics, although somewhat earlier, in the 1620s, plague nearly destroyed the Jamestown colony. Likewise, after the colonial period, America rarely experienced bubonic plague outbreaks, except during worldwide epidemics.

Do you recall receiving, in 1769, a letter about actuarial principles from your English friend, the clergyman Richard Price? The communication contained this observation that, in light of present knowledge, proved remarkably prescient: "Among the peculiar evils to which great towns are subject, I might further mention the PLAGUE. Before the year 1666 this dreadful calamity laid London almost waste once in every 15 or 20 years...A most happy alteration has taken place, which, perhaps, in part, is owing to the greater advantages of cleanliness and openness, which London has enjoyed since it was rebuilt [after the Great Fire of 1666], and which lately have been very wisely improved."

Price attributed the absence of plague epidemics after 1666 to a more open and cleaner London following the fire. He was correct, although it took mankind 130 years to realize clean cities, by decreasing rat populations, reduced the risk of epidemic.

From: "Stuart Green" <stuartgreenmd@yahoo.com>
To: "Benjamin Franklin" <dr_benjamin_franklin@yahoo.com>
Subject: **Take one ounce of good Peruvian bark**

Dear Doctor Franklin:

In your time, **malaria** (ague or intermittent fever in your vocabulary) was a seasonal disease thought carried by miasmic vapors. More than fifty years after your death, researchers discovered that the malevolent mists emanating from swamps were, in fact, clouds of mosquitoes.

Dr. Ronald Ross, a British military physician working in India at the end of the 19th century, discovered that mosquitoes transmitted malaria from one person to another during their blood meals, the first such observation regarding any insect transmitted disease. By then, microscopists had already determined that a tiny parasite caused malaria.

Ronald Ross

Usually a person survives the first infestation of the malarial parasites (characterized by high fevers and chill, severe headaches, nausea, vomiting, delirium and sometimes, intestinal pains) and recovers, seemingly normal thereafter. Unfortunately, the microbes survive, migrate to the victim's liver, increase their numbers and eventually re-enter the bloodstream to cause another feverish attack.

Modern medicines derived from the bark of a South American Cinchona tree kill the malarial parasites, usually curing the disease.

You were obviously impressed by the effectiveness of Peruvian (or Jesuit) bark, since you took it so often. Your 1750 *Poor Richard's Improved* offered this advice for malaria: "Take one ounce of good Peruvian bark," and mix it with "treacle or molasses," to mask bark extract's bitter taste. You also suggested adding "twenty or thirty drops of laudanum [opium]…washing it down with a glass of Madeira or red wine." (Your concoction would make anyone feel better.)

You recommended bark to so many people that one would think you imported the product yourself! Here's an example written to your friend Reverend Samuel Johnson (soon to become President of King's College, now called *Columbia University*): "I am sorry to hear of your illness: If you have not been used to the fever…let me give you one caution. Don't imagine yourself thoroughly cured, and so omit the use of the bark too soon. Remember to take the preventing doses faithfully."

Cupping

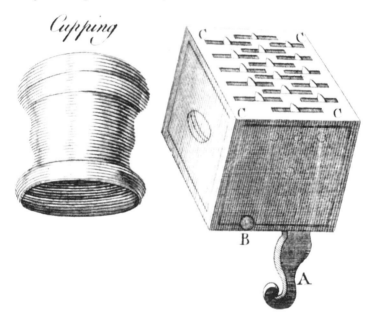

18th-century cupping and bleeding instruments

Do you remember becoming ill in 1757 while in London? [If memory serves, you were there to convince the Penn family to pay a share of Pennsylvania's taxes.] You sent a letter home to your wife Deborah informing her that you "had a violent cold and something of a fever, but that it was almost gone." The illness might have been a bout of malaria because you told Deborah, "it was not long before I had another severe cold, which continued longer than the first, attended by great pain in my head, the top of which was very hot…" Recall that your headaches "continued sometimes longer than at others; seldom less than 12 hours, and once 36 hours." Moreover, you reported that you were "now and then a little delirious." For treatment, recall being "cupped" on the back of the head (a suction treatment still popular in China).

You also mentioned that "I took a great deal of bark both in substance and infusion, and too soon thinking myself well." You soon relapsed: "I ventured out twice, to do a little business and forward the service I am engaged in, and both times got fresh cold and fell down again; my good doctor grew very angry with me, for acting so contrary to his cautions and directions, and obliged me to promise more observance for the future."

As for medicine, you told Deborah, "I took so much bark in various ways that I began to abhor it." Finally, you reported "I was seized one morning with a vomiting and purging, the latter of which continued the greater part of the day, and I believe was a kind of crisis to the distemper, carrying it clear off; for ever since I feel quite lightsome, and am every day gathering strength."

Chinchona tree

In spite of the unpleasant effects of bark, you later recommended it to Deborah: "I hope that very bad cold you had is gone off without any ill consequences. I have found by a good deal of experience, that three or four doses of bark taken on the first symptoms of a cold, will generally put it by."

Several years later, your grandson William Temple Franklin wrote to you, "I have now the pleasure of informing you, (and indeed it is a great one) that I have got the better of that stubborn monster the fever…to the red Peruvian bark, that I am indebted for this almost immediate relief."

You came up with a clever response when you wrote back, "You are well advised to continue taking the bark. There is an English proverb that says, *an ounce of prevention is worth a pound of cure*. It is particularly true with regard to the bark…"

The best story I've come across about the benefit of Cinchona bark appeared in an early issue of *Poor Richard's*: When the son of the French King Louis XIV fell ill with a fever no physician could heal, he called for "Talbot, an illiterate Englishman," who had grown famous for curing maladies with secret medications. Talbot confronted the Royal doctors when he told them that the future King was suffering from "ague—'Tis a distemper-that I can cure, and you can't."

The physicians warned Louis XIV that Talbot was "an ignorant quack, and not fit to be trusted with the Dauphin's health." The King nevertheless went along with Talbot's remedy, which proved successful. Louis XIV purchased a quantity of Talbot's medicine "at a great price for the publick good, and it proved no other thing than the bark disguised."

Your final comment has proven prophetic: "'Tis not unlikely, that some other valuable old medicines have been disused, from like causes, and may in time be advantageously revived again, to the benefit of mankind."

How right you were, Dr. Franklin. Today's researchers scour old medical books looking for discarded remedies that might contain some ingredient "to the benefit of mankind." They also realized that the best medicines to fight infections come from nature itself. But more about that in a future email. Tomorrow, I'll email you about another frequent visitor to colonial Philadelphia—yellow fever.

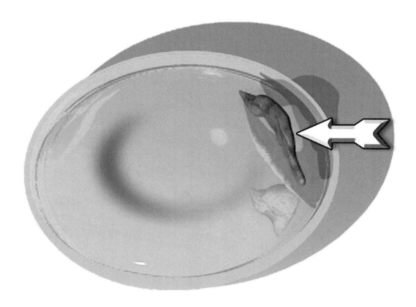

Malaria parasite (arrow) inside red blood cell

From: "Stuart Green" <stuartgreenmd@yahoo.com >
To: "Benjamin Franklin" <dr_benjamin_franklin@yahoo.com>
Subject: **The yellow or bilious fever**

Dear Doctor Franklin:

Recall that Philadelphia experienced two **yellow fever** epidemics while you lived there, the first in 1741 and again in 1762. You obviously queried medical friends about the disease because their responses are among your papers.

Your friend, the Virginia physician John Mitchell, wrote you about yellow fever in 1745. He recommended purging to cleanse the bowels "of their feculent and corruptible contents."

Thomas Moffatt, the prominent Connecticut physician, reminded you of the well-known seasonal nature of yellow fever (although never making the connection to mosquitoes) when he wrote, "the yellow or bilious fever …being the confessed offspring of heat and moisture may be brought into Philadelphia in your hot season…" However, he considered Philadelphia's climate too mild "to generate and produce it and therefore will decline or disappear with the first temperate and cool weather."

Three years after your *embarilment-vivant*, yellow fever hit Philadelphia hard. Nearly ten percent of the population succumbed within three months. Many physicians threw up their hands in defeat, accepting the clergy's explanation of Divine judgment, although your protégée Dr. Benjamin Rush took a different perspective. He considered such rationalizations archaic. He held instead to the miasma notion of spread, coupled with a humoral imbalance outlook. Thus, he advocated bloodletting and purges, a position he defended to the end of his life.

Rush in his *Account of the Philadelphia Epidemic of 1793,* never recognized that the epidemic came to Philadelphia with hundreds of white plantation owners fleeing Haiti after a slave uprising. He did, interestingly enough, acknowledge the high number of mosquitoes during the epidemic but said they were responding to the disease-causing vapors.

Shortly after Ronald Ross's 1900 discovery that mosquitoes transmit malaria, U.S. Army physician Walter Reed led a group of doctors to Cuba to deal with a yellow fever epidemic. They proved that mosquitoes spread the disease. Unlike Dr. Ross, however, Reed's group never actually saw the causative germ, a submicroscopic virus.

Today, we have a costly vaccine for yellow fever, given to travelers planning trips to tropical jungles.

Ronald Ross and Walter Reed assumed that their discoveries would rid the world of something undesirable, which they indeed accomplished: the miasmists soon disappeared! Malaria and yellow fever, however, remain—particularly in poor countries lacking the capital to drain swamps as your Council did in Philadelphia.

In the next email, I'll tell you about a most fortuitous discovery by a British researcher named Alexander Fleming that did much to conquer contagious diseases throughout the world.

Alexander Fleming

From: "Stuart Green" <stuartgreenmd@yahoo.com>
To: "Benjamin Franklin" <dr_benjamin_franklin@yahoo.com>
Subject: **A continual risque to my health**

Dear Doctor Franklin:

In a previous email I told you about the origin of colchicine for gout. Now, as promised, I'll acquaint you with the medicine for your lung infection.

During the century after yours, microscopists and biologists gradually identified the tiny animalcules that cause empyema, boils, and other suppurations. These *bacteria* exist in a myriad of shapes and species. Round germs in clusters and chains bring about boils and pneumonias; rectangular ones give rise to anthrax.

Spiral bacteria produce the disease you worried about acquiring by "that hard-to-be-governed passion of youth, had hurried me frequently into intrigues with low women that fell in my way, which were attended with some expence and great inconvenience, besides a continual risque to my health by a distemper which of all things I dreaded, tho' by great good luck I escaped it."

Although boils, empyema, anthrax, syphilis and plague differ, they share one feature in common: the same medicine cures them all: *penicillin*, discovered quite by accident at London's St. Mary's Hospital in 1929. By then, scientists had learned how to cultivate bacteria for analysis using a seaweed product you'd call *gelatina*—a congealed juice—but we call *agar*.

Alexander Fleming, a careless laboratory worker, left some of his bacteria-containing jars laying about uncovered and soon noticed that mold had grown on the agar's surface. The mold appeared as in your MICROSCOPE article, "extremely small, but perfect plants." You correctly surmised that "the seeds of these minute plants must, in all probability, be diffused universally through the air, and falling upon substances fit for their growth, spring up in astonishing profusion," for that's how the mold got on the agar.

Fleming made a startling discovery: the *Penicillium notatum* mold produced a substance that poisoned bacteria but proved harmless to people.

After penicillin's introduction into medical practice, doctors found that it didn't cure all bacterial diseases, so they started looking for similar substances that might destroy the resistant germs.

Penicillin's effect on bacteria occurred naturally in soils. There, bacteria and moulds compete with each other for nutrition. In a sense, you correctly surmised the situation when you wrote concerning population growth, "There is, in short, no bound to the prolific nature of plants or animals, but what is made by their crowding and interfering with each other's means of subsistence."

In the battle for soil dominance, Dr. Franklin, bacteria have the advantage of rapid growth. Nature, it turns out, gave something to mold as well—a substance to inhibit bacterial expansion.

For this reason, researchers test soils from around the world for molds that produce bacteria-killing *antibiotics*. Today, we now have dozens of such medicines, many produced by cultivating molds and similar microscopic beings. Antibiotics cure many contagious distempers, including tuberculosis, cholera, leprosy and plague. They also help prevent infections after surgery, the common complication of operations in your times.

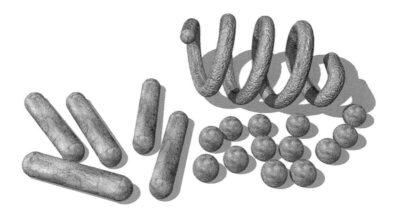

Types of bacteria killed by Fleming's Penicillin include the cause of anthrax (rods), boils (spheres), and the distemper spread by low women (spiral). (X10,000)

For this reason, you needn't fear bladder stone surgery when your doctors recommend it.

I believe, Sir, you anticipated just such improvements in health care, although without specific details, when you realized that you and your generation had created a new mode of progress where rational inquiry, laboratory experiments and statistical methods supplanted superstition and dogma.

In the three centuries since your birth, the average lifespan for Europeans, Americans and others of advanced nations has increased markedly from your times. Thus, living to the 84th year of life as you did before encaskment, is now common among us.

Sadly, certain backwards and still developing nations continue with short average life spans, due mostly to contagious diseases that carry off the young. The control of bacteria, you see, has not been as simple as I've described it; the germs change from one generation to the next in response to antibiotics and come to resist destruction in unique ways. How they do this—modify their structures—will be the subject of a future email.

From: "Stuart Green" <stuartgreenmd@yahoo.com >
To: "Benjamin Franklin" <dr_benjamin_franklin@yahoo.com>
Subject: **Great numbers of you catch cold**

Dear Doctor Franklin:

This email deals with a viral illness of particular interest to you—the common cold. Although today's remedies help with a runny nose, itchy eyes, sneezing, raspy throat and a dry cough, medical scientists haven't yet found a "cure" for the common cold.

The reason: more than 150 *different* viruses cause colds. Even though life-time immunity follows an infection by any one of the viruses, susceptibility to all the rest remains.

Because laboratory animals don't become infected with cold viruses, modern research requires volunteers to willingly inhale the germs to see what happens.

Before getting to what scientists have learned, I want to review your thoughts about colds for my own edification. Please correct me if I've mis-understood your ideas.

As with your analysis of any disease, your writings contemplate three dis-tinct issues: 1) the means of acquisition; 2) the nature of the illness; and 3) treatment strategies.

Like everyone before or since, you assumed that certain patterns of be-havior promoted colds. *Poor Richard* for instance, reproached readers, "you are apt...to expose yourselves too much and too long to the night air, whereby great numbers of you catch cold."

That wetness or coldness causes colds remains a common notion today. Recall that in the 1730s you held to a theory of pores, with both excreting skin pores to let out "the perspirable matter" and imbibing pores that drew in liquids. According to your earliest statements on the subject, if a person with part of their body warm—which opens the pores—is exposed to a sud-den cold draft, frigid air can enter the pores and cause a cold. As you said, "when before common fires, by which many catch cold, being scorcht be-fore and as it were froze behind."

Moreover, you supported your conclusion about drafts by quoting from a Chinese treatise, *The Art of Procuring Health and Long Life*: "of all the malignant affections of the air, a wind that comes thro' any narrow passage, which is cold and piercing, is most dangerous...causing grievous diseases." Such drafts should be avoided, according to an ancient Chinese proverb, "as carefully as the point of an arrow."

Absent a draft, you said, warm or cold air won't, by itself, cause a cold. Here's how you put it: "That warm rooms made people tender and apt to catch cold is a mistake." You cited as examples, our ability to "leap out of the warmest bed naked in the coldest morning, without any such danger; and in the same manner out of warm clothes into a cold bed" without catching a cold. You searched the world to muster support for your claim that abrupt changes in exposure to heat or cold are not harmful: "the Swedes, the Danes, the Russians" who, you said, lived in rooms "as hot as ovens" yet were particularly tolerant of the cold.

Having lived for a while in Siberia, I confirm what you said about the Russians' remarkable pastime: It was surprising "how far the Russians can endure heat; and how, when it makes them ready to faint, they can go out of their stoves [we call them *saunas*], stark naked, both men and women, and throw themselves into cold water; and even in winter wallow in the snow."

You also debunked the belief that moisture caused colds. In 1773 you perceptively wrote, "The gentry of England are remarkably afraid of moisture, and of air. But seamen who live in perpetually moist air, are always healthy if they have good provisions." After mentioning that people on tropic islands constantly exposed to sea spray remain perfectly healthy, you offered this opinion: "I have long thought that mere moist air has no ill effect on the constitution."

Benjamin Rush

In a 1773 letter you came closer to the truth than even you realized: "I am persuaded we are on a wrong scent in supposing moist, or cold air, the causes of that disorder we call a cold. Some unknown quality in the air may perhaps sometimes produce colds, as in the influenza."

How right you were, Dr. Franklin, as I'll explain in a moment.

In your correspondence with friends, especially after 1750, you shifted towards a contagion viewpoint about the origin of colds: "Many London families go out once a day to take the air; three or four persons in a coach, one perhaps sick; these go three or four miles or as many turns in Hide Park, with the glasses both up close, all breathing over and over again the same air they brought out of town with them in the coach with the least change possible, and rendered worse and worse every moment."

William Cullen

You found it amusing that the British called this practice "taking the air."

Do you remember answering Dr. Benjamin Rush's inquiry about your theory of colds? He solicited your opinion about a book by the Edinburgh physician William Cullen suggesting that colds were contagious illnesses. You told Rush you were glad Cullen "speaks of catarrhs or colds by contagion." After that, you made a particularly farsighted statement: "I have long been satisfied from observation, that...people often catch cold from one another when shut up together in small close rooms, coaches, &c. and when sitting near and conversing so as to breathe in each others transpiration..."

You also suspected that "perspired matter from our bodies" might contain a putrid substance that stays in bedclothes, clothing and even books that have not been cleaned after exposure to a sick person. These items can "infects us, and occasions the colds" after sleeping in unclean beds or wearing contaminated clothes.

Your observation, mixed with a good dose of common sense, resulted in an improved theory of colds more consistent with modern research than the beliefs of many doctors of your times.

Cold viruses X1,000,000

By 1788, you were clearly more enlightened than Dr. Rush about the spread of colds. He held fast to the traditional view that "in the histories of all epidemics, whether plague, small pox, putrid fevers &c. we find the operation of cold and moisture in the most sensible manner upon the body either in predisposing to, preventing or changing the type of these diseases." He also made the following erroneous pronouncement: "But I go further; we find several diseases actually produced by cold such as the rheumatism, angina, and pleurisy."

You must have felt odd, thinking that your own opinion of colds (as a contagious disease) were so "singular" in London that you were almost afraid to tell anyone about them. You shouldn't have felt that way, Sir, because time has proven you correct.

From: "Stuart Green" <stuartgreenmd@yahoo.com>
To: "Benjamin Franklin" <dr_benjamin_franklin@yahoo.com>
Subject: **A disgrace among the ancient Persians to cough or spit**

Dear Doctor Franklin:

With the advent of *microbiology* in the mid 19th century based on the discoveries of Pasteur, Koch and others, the obvious contagious nature of colds and influenza led physicians and scientists to hunt for causative germs. Researchers expected an early victory, but it didn't happen.

A common cold germ eluded detection, but not for a shortage of candidates. The nose and throat proved full of bacteria; microscopists have identified numerous species, but none caused colds when transferred to volunteers.

As the 20th century opened, the failure to find a common cold germ led some physicians to revive theories from your era. They blamed chilling, especially going from a warm room into the cold air.

The answer to the riddle of colds came from Germany. In 1914 Dr. Walter Kruse, a follower of Koch, blew his nose during a cold and, after filtering out the bacteria, inoculated the remaining fluid into the noses of volunteers. Kruse discovered that viruses—tiny enough to pass through a paper filter—caused colds.

Viruses are so small that if puffed-up to the size of a plum, a human enlarged to the same scale would have his head in Maine and his feet in Florida. Because they must reproduce in living cells, scientists developed techniques to grow viruses in the fluid surrounding chick embryos still in their eggs.

Scientists have concluded that wet feet, drafts and damp clothing don't cause colds, just as you said. Exposure to a person expelling cold viruses during a cough or sneeze is what really matters.

Even though you correctly judged colds as contagions, when you caught a particularly serious cold while in London, you dutifully submitted to your doctor's treatment, including Peruvian bark and bleedings.

In a previous email, I mentioned an illness afflicting you a few months after arrival in London in 1757, for which you consumed large quantities of Peruvian bark.

The clinical course of that malady, in my opinion, had symptoms not ordinarily associated with a common cold or malaria. I suspect that your intestinal complaints came from the quinine (bark). You described it yourself thus: "I took so much bark in various ways that I began to abhor it."

Here, Dr. Franklin, are common *side effects* listed on a bottle of modern quinine tablets: "abdominal or stomach cramps or pain; blurred vision;

change in color vision; diarrhea; headache (severe); nausea or vomiting; ringing or buzzing in ears or loss of hearing."

Three years later, still in London, you wrote Deborah that you were "much indisposed with an epidemical cold" associated with a headache. Dr. Fothergill, you wrote, drained "8 ounces of blood from the back of my head." Since you didn't send the letter off right away, you had the opportunity to write this postscript: "I was blooded on Sunday, 16 ounces, which was of great service; but that and physic has left me a little weak."

John Fothergill

That Fothergill bled the back of your head, rather than your forearm, was entirely consistent with then current practices. Phlegm, one of the four humors, leaked from your nose during the catarrh, just as water flows over a cataract (a word from the same root). Since doctors knew that a watery fluid surrounded the brain, it seemed reasonable that the same fluid dripped out of the nose.

My key to understanding your thoughts on cold—and why bleeding from the head seemed appropriate—are found in the *Notes on Colds* you wrote to yourself in 1773. It's too bad you never completed *Notes* because some are difficult to decipher.

While I realize you often made lists of ideas for future discourse and publication, by leaving the project unfinished, you've generated unending speculation about the meaning of a word here or a phrase there. Nevertheless, *Notes* contained many intriguing concepts.

You started by describing a cold as a "…thickness of the blood, whereby the smaller vessels are obstructed, and the perspirable matter retained…" This thickened blood "overfills the vessels" and "produces coughs and sneezing by irritation."

Blockage of sweating (by either dried sweat or closure of pores by cold air) has the same effect: "diminished perspiration," which thickens the blood. Thick blood, in turn, causes fluid to flow out the nose. Likewise, gluttonous eating and drinking overwhelms the body's ability to sweat out the excess, thickening the blood.

You also thought one could retain too much sweat "by cooling suddenly in the air after exercise." You explained this seeming paradox: "Exercise quickening the circulation, produces more perspirable matter in a given time, than is produced in rest." Suddenly quitting exercise retains some perspiration before all of it is expelled, implying that we should gradually cool down towards the end of a workout, rather than stopping suddenly.

(Today, all exercise regimens involve a cooling down period, although not to prevent colds.)

It was clever how you described practices from around the world that supported your theory of colds: "American Indians, in the woods, and the whites in Imitation of them, lie with their feet to the fire in frosty nights on the ground, and take no cold while they can keep their feet warm."

Most significantly from a modern perspective, you also said, "It was a disgrace among the ancient Persians to cough or spit. Probably as it argued intemperance." This sentence, more than any other, suggests that you realized phlegm and sputum transferred colds from one person to another.

I agree with your statement that it would be "A general service to redeem people from the slavish fear of getting cold by showing them where the danger is not and that where it is…" Not much has changed since you wrote those words.

The only cure for a cold is tincture of time. Medicines taken for cold symptoms may actually perpetuate the illness by reducing natural immunity. People, however, still cling to ancient notions about the cause and treatment of colds. Likewise, some of our present remedies are as effective curing colds as was bloodletting in your time.

My next group of emails, Sir, will review progress in the spark of your great fame—electricity.

Louis Pasteur

Robert Koch

ELECTRICITY

✉ Produce something for the common
benefit of mankind 81

✉ The electrical matter consists of particles
very subtile 85

✉ Secured from the stroke of lightening 87

✉ Mischief by thunder and lightning 91

✉ To the magazines at Purfleet 95

✉ *Erepuit coelo fulmen sceptrumque tyrannis* 97

✉ Electricity in palsies 99

✉ M. Volta's experiment 103

✉ Electricity and magnetism 107

✉ The conducting quality of some
kinds of charcoal 111

✉ Some principle yet unknown to us 113

✉ As thro' a vacuum 115

18th-century electricity experiments

From: "Stuart Green" <stuartgreenmd@yahoo.com >
To: "Benjamin Franklin" <dr_benjamin_franklin@yahoo.com>
Subject: **Produce something for the common benefit of mankind**

Dear Doctor Franklin:

Although you spent only five years performing electrical experiments, you'll be pleased to learn that certain of your proposals, conjectures, and conclusions were at least a century and a half ahead of their time. A scientist named Robert A. Millikan won a prestigious award in 1923 called the *Nobel Prize in Physics* for devising a simple yet elegant experiment to measure the charge carried by electrons–the individual particles of your "electric fluid." He thus carried forward the research that gave you widespread fame during your lifetime.

Millikan wrote about you: "within two years of the time of his first experiment had acquired a keener insight into the fundamental nature of electrical phenomena, not merely than any had acquired up to his time, but even than any one of his successors acquired for the next hundred and fifty years, when, about 1900, the scientific world returned essentially to Franklin's views."

The road upon which your ideas traveled, however, was strewn with bumps and obstacles, some placed there by others and some of your own making. Your first theory of electric charge, for instance, contained errors that you wisely modified after further experiments. Indeed, today's scientists appreciate your candid statement that, "If there is no other use discovered of electricity, this, however, is something considerable, that it may help to make a vain man humble."

To be honest, Dr. Franklin, historians debate the priority of certain ideas generally attributed to you, with a few experts ascribing your conclusions to contemporaries. As you know, Sir, the process of challenging your originality started while you lived. At least one European, the Czech prelate Procopius Divis, claimed *he* was the first person to erect a grounded metal rod to control lightning strikes, an assertion ridiculed by most modern historians.

One of the most contentious debates about your experiments, however, has lasted more than two hundred years. Several scholars have argued that you *never* flew a kite in an electric storm and didn't erect lightning rods in Philadelphia until *after* they went up in Europe. With a sentence or two, Dr. Franklin, you could end such disputes. For now, the best information we have comes from Joseph Priestley's description of your kite experiment he claimed he got "from the best authority" (presumably you). Priestley wrote:

The Doctor…was waiting for the erection of a spire [on Christ Church] in Philadelphia to carry his views into execution…when it occurred to him that by means of a common kite he could have better access to the regions of thunder than by any spire whatever...But, dreading the ridicule which too commonly attends unsuccessful attempts in science he communicated his intended experiment to nobody but his son...who assisted him in raising the kite.

The kite being raised, a considerable time lapsed before there was any appearance of its being electrified...Just as he was beginning to despair of his contrivance, he observed some loose threads of the hempen string to stand erect...Struck with this promising appearance, he immediately presented his knuckle to the key, and (let the reader judge of the exquisite pleasure he must have felt at that moment) the discovery was complete.

I hope you'll email me the *real* story of your kite experiment if different from the above. Meanwhile, I'll review the features of electricity that philosophers of your era found so interesting.

In the centuries before your time, people feared peculiar natural occurrences they could neither understand nor even relate to each other. Both Greek and Roman fishermen, for instance, knew that the torpedo fish gave its captor a painful shock, which momentarily paralyzed muscles and caused an uncomfortable tingling sensation in the limbs. They realized that a wet fishing line could convey the jolt without even handling the fish.

Second, lightning evoked fear whenever a storm threatened our mutual ancestors, who sought theological justification after every fatality or property loss. Ancient philosophers assumed that a fire in the clouds accounted for lightning's strange features.

Lastly, one of the most puzzling features of electricity completely defied comprehension: the ability of amber, when rubbed, to attract bits of wheat chaff and dry grass, whereas the lodestone, a naturally-occurring magnet, drew only iron to itself and didn't require rubbing to achieve the effect. (Moreover, when suspended by a string, such a rock would align itself along a north-south axis and could permanently confer geological polarity to an iron needle.)

In 1629, Niccolo Cabbeo—a Jesuit priest and scientific researcher—observed that some substances, after first being drawn to rubbed amber, flew off the stone's surface. He misinterpreted what he observed as a rebound phenomenon, with the attracted particles bouncing off the amber because they were, at first, drawn to it so strongly. Nevertheless, history credits Cabbeo as the discoverer of electric repulsion.

In the year you were born, Dr. Franklin, the English scientist Francis Hauksbee, while setting up an electricity demonstration at the Royal Society of London, noticed that when he shook a vacuum tube containing mercury, a faint blue glow developed within. Hauksbee suspected that the light effect might be electrical, so he put mercury in a glass vacuum tube on a spindle-and-crank assembly. By rubbing the rotating glass, Hauksbee stimulated a glow bright enough to read by. He had made an *electrostatic* generator, stronger than one created from a rotating sulfur sphere years earlier in Germany.

Stephen Gray, another Englishman interested in scientific research, became fascinated with Hauksbee's electricity generator. A cork placed in the *end* of a charged tube, he discovered, attracted a feather with greater strength than the glass itself. Gray tried a number of different substances and soon found that iron or brass wire can conduct the "electric virtue" over significant distances.

Curious about how far the electric charge would travel, Gray in 1729 hung a metal wire from silk insulating threads and found that the electric virtue could travel as much as 650 feet to the wire's end where an attached ivory ball attracted metal foil.

Later, French researcher Charles DuFay learned that gold leaf, after being drawn to an electrified glass tube, would jump away. But after the gold fell on another object, it jumped back to the glass, oscillating on and off until the tube's charge dissipated. DuFay shared his observations with Gray. They both concluded that "electricity" is a fluid that is transferable from one object to another just as water poured from a first glass into a second.

Gray suspected that one could collect electric fluid, which he related to thunder and lightning. He coined the term "electric fire" because a spark accompanied the transfer. He also puzzled over the way iron rods conveyed electricity: A pointed rod behaved differently from a rod with a blunt tip.

Recall that DuFay eventually concluded that two kinds of electricity existed: *resinous* electricity (created by friction against amber), and *vitreous* electricity (produced when glass is rubbed). You later gained fame by refuting his "two fluid" concept.

Since Gray knew that the human body could conduct an electric charge, he suspended a boy from silk cords and found that by rubbing the lad's foot against a rotating glass tube, his face would attract metal foil.

It soon became fashionable to hold public electricity demonstrations, complete with electric generators and a young lad suspended by silk ropes. In Germany, with typical continental flair, a pretty girl on a swing served in place of Gray's boy. Anyone kissing the lass received a shock—rather dramatic in a darkened room.

Do you remember Pieter van Musschenbroek, professor of experimental physics at Leyden? (How could you forget him? In 1759 he sent you two dozen books about electricity.) In the early 1740s, van Musschenbroek transferred a charge from a generator into a jar of water via a brass wire. While holding the jar in one hand, he touched the wire with the other and got the shock of his life. As he reported to the French Academy, "...I advise you never to try it yourself, nor would I, who have experienced it and survived by the Grace of God, do it again for all the Kingdom of France."

Van Musschenbroek, you'll recall, invented the Leyden jar (we now call it a *capacitor* or *condenser*) capable of accumulating and storing an electrical charge. Soon, experimenters shocked many people simultaneously with a single jolt from a Leyden jar. A whole row of soldiers were brought before the King of France, each connected to the next by an iron wire. They all jumped in unison from a Leyden jar discharge, greatly amusing the King.

At this point in the history of electricity, your principle British rival for priority of electric discoveries, William Watson, began his studies of electric transmission. He tried to determine electricity's speed by sending a charge along the Thames River, but found the transit too fast to measure. Watson realized that electric charges exhibited a polarity, which he indicated as + and - signs, a concept erroneously attributed to you by posterity (an error I hope you'll acknowledge). Watson also lined both the inner and outer surfaces of a Leyden jar with silver foil greatly increasing its charge-storing capacity.

Meanwhile, French experimenters connected several Leyden jars together in parallel, creating quite a charge. Dufay's former student Abbé Nollet killed a sparrow with such a jolt and had the bird autopsied. The physicians told him that the animal's organs resembled those of a person struck by lightning.

If you've any memory left, Dr. Franklin, you'll no doubt recollect the electricity demonstration that changed your life. In 1743, you attended Scottish physician Archibold Spencer's show in Boston. He electrified a suspended boy's feet with a glass generator and showed how bits of metal foil were drawn to his face. The following year, with your sponsorship, Spencer put on an electricity show in Philadelphia, creating quite a sensation.

Shortly after you saw Spencer's Boston event, Sir, your London book-purchasing agent Peter Collinson sent you magazine reports on German electrical experiments and a long glass tube designed for electricity research. You quickly began conducting experiments—studies that at first led you into conflict with some of the most powerful men in English science but ultimately brought you great esteem.

I envy the way you retired from business in the late 1740s, taking, as you put it, "the proper measures for obtaining leisure to enjoy life and my friends…having put my printing house under the care of my partner David Hall, absolutely left off bookselling, and removed to a more quiet part of the town, where I am settling my old accounts and hope soon to be quite a master of my own time, and no longer (as the song has it) at every one's call but my own."

I think you chose wisely when you declined to run for election to the Pennsylvania Assembly and even mentioned that you'd "not serve if chosen." I doubt you ever imagined, however, the consequences that followed your commitment to enjoy "leisure to read, study, make experiments, and converse at large with such ingenious and worthy men as are pleased to honour me with their friendship or acquaintance, on such points as may produce something for the common benefit of mankind, uninterrupted by the little cares and fatigues of business."

Your decision favorably influenced both science and world history.

From: "Stuart Green" <stuartgreenmd@yahoo.com>
To: "Benjamin Franklin" <dr_benjamin_franklin@yahoo.com>
Subject: **The electrical matter consists of particles very subtile**

Dear Doctor Franklin:

It's clear that once you began your electricity research, you became obsessed with the subject, devoting great energy to experiments—often with the help of friends Phillip Syng, Thomas Hopkinson and Ebenezer Kinnersley. Do you recollect telling Collinson in 1747, "I never was before engaged in any study that so totally engrossed my attention and my time as this has lately done; for what with making experiments when I can be alone, and repeating them to my friends and acquaintance, who, from the novelty of the thing, come continually in crowds to see them, I have, during some months past, had little leisure for anything else"?

In that same letter to Collinson, you wrote that you had been "making electrical experiments, in which we have observed some particular phenomena that we look upon to be new. I shall, therefore communicate them to you in my next [letter], though possibly they may not be new to you, as among the numbers daily employed in those experiments on your side the water, 'tis probable some one or other has hit on the same observations."

Well, Sir, researchers in Britain did *not* hit upon your observations.

Collinson reported your progress at meetings of the Royal Society of London. At first, the communications served as a source of mirth among the skeptical members, with electricity researcher William Watson leading the merriment. (Eventually though, your work won them over.)

Anyone doing electrical experiments at the time knew that a person insulated from the ground by standing on wax could receive an electrostatic charge from a generator and then transfer that charge to a grounded person by contact.

You ingeniously went further: You had *two* people stand on wax, with a third person—yourself?—on the ground.

You correctly concluded from this and other experiments that electricity was conserved in the same way Newton's momentum was conserved. The difference in the way two people on wax discharged an electric spark depended on the presence of either an excess or deficiency of electric charge. Moreover, the ground was a great repository that could either absorb excessive charge or supply a deficiency.

Correct me if I'm wrong, but as I understand it, each of the three subjects in the experiment, "has his equal share before any operation is begun with the tube." In other words, all three people started neutral in their electric charge. The spinning glass generator drew electricity from the person who touched the tube. Since that person was insulated from the ground, he now had a deficiency of electricity and the tube possessed an excess amount. When the second person on wax touched the now stationary tube, some of the tube's excess charge flowed to that person until the second person and the tube were in equilibrium—both with an excess charge

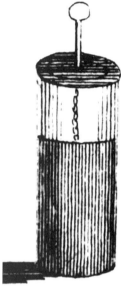

Leyden Jar

If the grounded person touches *either* of the two people standing on wax, he experiences a shock. To the first person on wax, the grounded subject *supplies* electricity *from* the earth by way of a spark, thus neutralizing that person's deficiency. To the second person on wax, the grounded subject *withdraws* the excess electricity *to* the earth, also feeling a shock.

As with the people on wax, the Leyden jar, you realized, could hold within either an excess or deficiency of charge, with the opposite situation on its exterior surface. In this way, the total quantity of electric charge was conserved—only the inside-to-outside distribution was unequal. Other scientists, the Abbé Nollet, for instance, familiar with the way a Leyden jar discharged, wrongly assumed that the electric matter flowed through pores in the glass from the inside to the outside, causing the imbalance.

You grasped a fundamental truth of nature, Dr. Franklin, when you concluded that electricity, rather than involving an imponderable substance, actually consisted of real particulate matter. As you shrewdly put it: "The electrical matter consists of particles extremely subtle, since it can penetrate common matter, even the densest metals, with such ease and freedom as not to receive any perceptible resistance."

I enjoyed reading about your amusing fake spider illusion based on the ability of electrified materials to display alternating movements. You wrote, "We suspend by a fine silk thread a counterfeit spider, made of a small piece of burnt cork, with legs of linen thread, and a grain or two of lead stuck in to give him more weight." The fake spider jumped back and forth from the wire of an electrified Leyden jar to a grounding wire, "playing with his legs against both in a very extraordinary manner, appearing perfectly alive to persons unacquainted." You told Collinson, "He will continue this motion an hour or more in dry weather."

In the same letter to describing the spider demonstration, you reported an observation of great historic significance. I'll get to it in my next email. For now, I've got to go.

From: "Stuart Green" <stuartgreenmd@yahoo.com>
To: "Benjamin Franklin" <dr_benjamin_franklin@yahoo.com>
Subject: **Secured from the stroke of lightening**

Dear Doctor Franklin:

In today's email, I'll review your most famous proposal, so please pay close attention. During your electricity research, you discovered "the wonderful effect of pointed bodies, both in *drawing off* and *throwing off* the electrical fire." (Something about the sharpness of a metal point made it more effective than a blunt rod in drawing off electricity.)

I assume you also recognized that the electrical effect involved *grounding* the charge because you could eliminate a metal point's neutralizing ability by insulating it from the ground.

Nobody knows exactly when you came up with the lightning rod because your letter to Collinson about the usefulness of pointed metal rods no longer exists. Luckily for posterity, *Gentleman's Magazine* published this reprint in 1750: "For the doctrine of points is very curious, and the effects of them truly wonderful; and, from what I have observed on experiments, I am of opinion, that houses, ships, and even towns and churches may be effectually secured from the stroke of lightening by their means."

You proposed that "on the tops of the weathercocks, vanes or spindles of churches, spires or masts, there should be put a rod of iron 8 or 10 feet in length, sharpened gradually to a point like a needle, and gilt to prevent rusting, or divided into a number of points, which would be better." You anticipated that with such a rod, "the electrical fire would, I think be drawn out of a cloud silently, before it could come near enough to strike."

I assume you intended to prove your lightning rod conjecture in the near future: "This may seem whimsical, but let it pass for the present, until I send the experiments at large."

In your earliest writings on lightning rods, it appears that you expected ungrounded metal points to draw electricity out of the air, thereby preventing lightning strikes. That was, of course, an incorrect conclusion, because a rod must be <u>grounded</u> to conduct the charge safely into the earth. You soon corrected your mistake and proposed grounding high-mounted points.

In 1755, a friend asked how you concluded that lightning was a form of electricity. Your response is justifiably famous:

I cannot but answer better than by giving you an extract from the minutes I used to keep of the experiments I made: ...Nov. 7. 1749. Electric fluid agrees with lightning in these particulars: 1. Giving light. 2. Colour of the light. 3.Crooked direction. 4. Swift motion. 5. Being conducted by metal. 6. Crack or noise in exploding. 7. Subsisting in water or ice. 8. Rending bodies it passes through. 9. Destroying animals. 10. Melting metals. 11. Firing inflammable substances. 12. Sulphureous smell. The electric fluid is attracted by points. We do not know whether this property is in lightning. But since they agree in all the particulars wherein we can already compare them, is it not probable that they agree likewise in this? Let the experiment be made."

Fig. 9

It's one thing to idly speculate about a lightning rod's effectiveness and quite another to propose an experiment to confirm the notion as you did: "To determine the question, whether the clouds that contain lightning are electrified or not, I would propose an experiment to be tried where it may be done conveniently. On the top of some high tower or steeple, place a kind of sentry box big enough to contain a man and an electrical stand."

We now call this your "sentry box experiment." Your instructions for bringing the electric charge into the box were not very precise, a number of modern commentators have pointed out, but I disagree.

What could be clearer than this? "From the middle of the stand let an iron rod rise, and pass bending out of the door, and then upright 20 or 30 feet, pointed very sharp at the end. If the electrical stand be kept clean and dry, a man standing on it when such clouds are passing low, might be electrified, and afford sparks, the rod drawing fire to him from the cloud."

I wonder why British researchers didn't rush to perform the sentry box experiment; after all, Peter Collinson (in April, 1751) published in Great Britain your book, *Experiments and Observations on Electricity, Made at Philadelphia in America, by Benjamin Franklin, and Communicated in several Letters to Mr. P. Collinson, of London, F.R.S.*, which contained the proposal. Indeed, we wonder today why you didn't build a sentry box yourself. Perhaps safety considerations were on your mind, having nearly electrocuted yourself while attempting to roast a turkey with electricity.

To your consternation, some British scientists didn't pay much attention to your book when it first appeared. You mentioned the matter in your *Autobiography*: "It was, however, some time before those papers were much taken notice of in England."

It's quite remarkable that French researchers tested your lightning-is-electricity conjecture first, a year after your book's publication. As you put it: "A copy [of your book] happening to fall into the hands of the Count de Buffon, a philosopher deservedly of great reputation in France, and indeed all over Europe he prevailed with M. Dalibard to translate them into French, and they were printed at Paris."

Your French admirers did more than simply publish your book; they actually constructed your proposed sentry box at Marly-le-ville. There, during a storm on May 10, 1752, your hypothesis about the equivalence of lightning and electricity was confirmed.

Certain rival electricians remained unimpressed. As you later wrote in your *Autobiography:* "The publication offended the Abbé Nollet, Preceptor in Natural Philosophy to the Royal Family, and an able experimenter, who had formed and published a Theory of Electricity, which then had the general vogue. He could not at first believe that such a work came from America, and said it must have been fabricated by his enemies at Paris, to decry his system."

After the sentry box event, you gained what you called "sudden and general celebrity." Parisians soon witnessed public demonstrations of your electrical experiments. Louis XV and his court were duly impressed.

Do you recall your reaction when news of French confirmation reached Philadelphia? As your *Autobiography* modestly puts it, "I will not swell this narrative with an account of that capital experiment, nor of the infinite pleasure I received in the success of a similar one I made soon after with a kite at Philadelphia, as both are to be found in histories of electricity." And surely, Sir, they are.

The British, embarrassed that French scientists proved you right before they did, quickly corrected their mistake. English scientists made amends by admitting you to the Royal Society without an annual fee and awarding you the prestigious Copley Medal in 1753.

Even William Watson acknowledged your talent: "Mr. Franklin appears in the work before us to be a very able and ingenious man…I think scarce any body is better acquainted with the subject of electricity than himself."

The Abbé Nollet, however, unlike his French colleagues, refused to accept your conjectures, especially the single fluid theory. As you stated, "he wrote and published a volume of letters, chiefly addressed to me, defending his theory, and denying the verity of my experiments and of the positions deduced from them."

Most researchers, faced with an attack on their conclusions, take the offensive. Posterity honors the way you avoided confronting Nollet, and made the following decision instead: "I concluded to let my papers shift for themselves; believing it was better to spend what time I could spare from public business in making new experiments, than in disputing about those already made. I therefore never answered M. Nollet; and the event gave me no cause to repent my silence…my book was…by degrees universally adopted by the philosophers of Europe in preference to that of the Abbé, so that he lived to see himself the last of his sect."

Sure enough, Dr. Franklin, you soon became elevated to near-Promethean status in your own lifetime, a position occupied by few men in all human history.

Most Americans today, so familiar with your involvement with our nation's independence, don't realize that your influence flowed from your scientific discoveries. I'm confident you'll agree that, but for your electrical experiments and conjectures, you might've retired from printing into a life of leisurely pursuits without gaining authority from international acclaim.

If your Madeira-soaked brain has remembered anything, Dr. Franklin, it's probably the way your fame spread after other scientists confirmed your experiments and conclusions. Nevertheless, I'll write about that adulation next.

Peter Collinson

From: "Stuart Green" <stuartgreenmd@yahoo.com >
To: "Benjamin Franklin" <dr_benjamin_franklin@yahoo.com>
Subject: **Mischief by thunder and lightning**

Dear Doctor Franklin:

We can only imagine today how pleased you felt when researchers confirmed your proposals about electricity. In 1752, the British tried to duplicate the sentry box trial but were foiled by the weather—good weather in this case. In that same year, the gods blessed France with sufficient thunderstorms to provide amply lightning for experimental purposes.

Elsewhere in Europe other electricians, you may recall, also got lucky with the weather during 1752 and 1753. One place in particular has become famous—I should say infamous—for its excellent lightning: St. Petersburg. On July 26th, 1753, the Swedish-born electrical experimenter Georg Richmann was electrocuted in his home by lightning conducted into his laboratory by his "thunder-machine" experiment. Richmann, as you know, inadvertently grounded his lightning rod through himself! He had already read your *Experiments and Observations*, which he ridiculed, claiming that he, Richmann, had actually discovered electric charges in the atmosphere.

In his death, Richmann did a great service to science. News of the Russian's electrocution spread rapidly through Europe's scientific circles.

Clearly you knew about the event by March 1754 when you published an account of Richmann's death and autopsy findings in your *Pennsylvania Gazette*. As you put it: "The new doctrine of lightning is, however, confirmed by this unhappy accident; and many lives may hereafter be saved by the practice it teaches…And had his apparatus been intended for the security of his house, and the wire (as in that case it ought to be) continued without interruption from the roof to the earth, it seems more than probable that the lightning would have followed the wire, and that neither the house nor any of the family would have been hurt by that unfortunate stroke."

Until your discovery, lightning remained the sole possession of the Deity, hurtled at the sinful for their transgressions. God reserved for himself the special power of lightning bolts to pinpoint destruction. It occurred to many clergymen that your lightning rod was somehow ungodly, so it doesn't surprise me that you enlisted Divine support for your ingenious invention.

By the time you received the Copley Medal, Dr. Franklin, you had already advocated lightning rods to protect homes and other structures from thunderbolts. Your 1753 *Pennsylvania Gazette*, to refresh your memory, contained instructions on *How to secure Houses, &c. from LIGHTNING*: "It had pleased God in his goodness to mankind, at length to discover to them the means of securing their habitations and other buildings from mischief by thunder and lightning." After describing how to erect the lightning rod, you told readers, "A house thus furnished will not be damaged by lightning, it being attracted by the points, and passing thro the metal into the ground without hurting any thing."

The spread of lightning rod technology in your time was slow but steady. In 1761, while in England as Pennsylvania's agent, you received from fellow Philadelphia electricity researcher Ebenezer Kinnersley a report that a lightning rod had recently protected the home of Philadelphia merchant William West. He said that West "suspected that the lightning, in one of the thunder storms last summer, had passed thro' the iron conductor which he had provided for the security of his house."

Kinnersley examined the rod and noticed melted metal in place of the thin wire that previously connected the rod to the ground. Moreover, one eyewitness saw the lightning strike the rod and another, according to Kinnersley, watched the charge travel from the rod's base along wet pavement into the ground.

Kinnersley ended with what became a common closing: "Sir, I most heartily congratulate you on the Pleasure you must have in finding your great and well-grounded Expectations so far fulfilled...May the Benefit thereof be diffused over the whole Globe. May it extend to the latest Posterity of Mankind; and make the Name of FRANKLIN like that of NEWTON, immortal." That gradually happened, Dr. Franklin, as I pray you'll recall.

Public acceptance of lightning rods didn't occur rapidly. In responding to Kinnersley, you wrote about the lack of lightning rods in London: "Here it is very little regarded; so little, that though it is now seven or eight years since it was made publick, I have not heard of a single house as yet attempted to be secured by it. It is true the mischiefs done by lightning are not so frequent here as with us, and those who calculate chances may perhaps find that not one death (or the destruction of one house) in a hundred thousand happens from that cause, and that therefore it is scarce worth while to be at any expence to guard against it"

As the value of lightning rods became evident, they sprouted on houses and buildings both in America and abroad. Peter Collinson, who supplied you with your first glass tube for electric experiments, and Dr. John Fothergill, the Quaker physician who helped get your letters published as *Experiments and Observations*, became effective spokesmen for lightning rods. Even Dr. William Watson acknowledged their importance.

Many clergymen, however, resisted lightning rods for theological reasons. They clung to the idea that the devil played a role in lightning strikes, best warded off by ringing church bells during storms. As the bells hung in steeples—usually a town's highest points—and because metallic crosses topped the steeples, many a bell-ringer got fried while chasing off evil spirits. With time, most churches eventually displayed well-grounded lightning rods affixed to their highest point.

You'll be particular pleased to learn that some lightning rod placements you personally devised still exist—and have continuously protected their structures from lightning damage. Do you recall sending plans for a lightning rod system to Maryland's Statehouse in Annapolis? Every day of the year, Sir, tour guides point to the grounded conductor atop the building's spire and proudly announce, "Benjamin Franklin designed it."

Maryland State House

Next, I want to briefly review for you the Purfleet controversy over pointed and blunt-tipped lightning rods, especially because it involves your strange friend Henry Cavendish.

Henry Cavendish

From: "Stuart Green" <stuartgreenmd@yahoo.com>
To: "Benjamin Franklin" <dr_benjamin_franklin@yahoo.com>
Subject: **To the magazines at Purfleet**

Dear Doctor Franklin:

In 1772, while still in England for your last diplomatic mission there, the Royal Board of Ordinance, you'll recall, asked you to serve on a commission to develop a way of preventing the Purfleet gunpowder magazines from exploding during a lightning strike. The Purfleet Committee included some of the finest minds in the United Kingdom. Henry Cavendish, the odd-mannered researcher I'll discuss in a later email, served on the committee alongside you. Do you recall the others? To refresh your memory: Dr. William Watson, your former electrical adversary; mathematician John Robertson, then Master of the Royal Naval Academy; and the painter Benjamin Wilson, yet another electrical experimenter who had earlier served with you on a committee to design a lightning rod system for St. Paul's Cathedral.

William Watson

(Do you remember sitting for Wilson's portrait? That painting now hangs in the *White House*, a structure built 10 years after your death as the official home of the America's President.)

The Committee's August, 1772 report stated that your group "visited those buildings, and examined, with care and attention, their situation, construction, and circumstances," and made recommendations to secure them from lightning.

I was surprised that only four members signed the document. Benjamin Wilson objected to the use of *pointed* lightning rods, claiming they would <u>draw</u> lightning from the sky. He proposed instead using blunt-tipped rods, which he thought safer.

Your own experiment with sharp and blunt tip conductors, each attached to an electroscope consisting of suspended balls that repel each other when charged.

I admire the sharp-tipped rebuttal to Wilson's charges you and your colleagues submitted on December 17th 1772: "Having heard and considered the objections to our report, concerning the fixing pointed Conductors to the Magazines at Purfleet, contained in a letter from Mr. Wilson...we find no reason to change our opinion, or vary from that Report." The pointed lightning rods were installed soon thereafter; no further gunpowder explosions occurred.

We now know that Wilson's dispute with you over pointed-verses-blunted lightning rods actually started with the St. Paul's project and never ended. As you told a correspondent, Wilson "was displeased that his opinion was not followed, and has written a pamphlet against points. I have not answered it, being against disputes." As we'll see in the next email, Wilson gained the King's ear—especially after America declared its independence.

From: "Stuart Green" <stuartgreenmd@yahoo.com >
To: "Benjamin Franklin" <dr_benjamin_franklin@yahoo.com>
Subject: ***Erepuit coelo fulmen sceptrumque tyrannis***

Dear Doctor Franklin:

People today, if asked what Benjamin Franklin did as a scientist, would respond, "he proved lightning is electricity." Few realize that your greatest claim to fame, especially in light of present knowledge, was your Single Fluid Theory of electric charge. That's because hardly anyone, save for historians of science, knows that before your time philosophers thought that two kinds of electricity existed: resinous electricity created by rubbing amber, and vitreous electricity produced while stroking glass.

As I confirmed in a prior email, you proposed that only one form of electricity exists. A charged object may have either a deficiency or an excess.

It was wise of you, Sir, to become so closely involved with the preparation of Joseph Priestley's 1767 book, *The History and Present State of Electricity, with original Experiments* because the text supported your side of the electricity story. By encouraging his own research, you also contributed to Priestley's development into a great scientist. (I'll remind you of his important work with gases later.)

By the time you arrived in Paris in late December 1776, you were hailed as *L' Ambassadeur Electronique*. In England, your role in fomenting America's independence made you the most hated of men by the monarch and his advisors. George III, in a fit of disdain, had pointed lightning devices atop Royal buildings changed to blunt-tipped rods, siding with those who considered the rounded conductors safer.

Your English scientist friends, however, continued to admire and correspond with you throughout the War of Independence. Many also spoke out against their country's inflexible attitude towards North America.

In France, meanwhile, Louis XVI, his court, and the general population loved your humble appearance and anti-British politics. No wonder the King's former finance minister, A.R.J. Turgot, said of you "*Erepuit coelo fulmen sceptrumque tyrannis*" because you *did* snatch lightning from the heavens and the scepter from a tyrant.

The new generation of French electrical researchers was called *Franklinistes* in your honor. To the modern ear, the names of these scientists—Coulomb, Ampère, Laplace, Fourier, and Poisson—are quite familiar. Some have units of electricity named after them, while others explained the precise mathematical relationship between electric charge and physical variables.

Poisson introduced in 1812 a new two fluid hypothesis for electricity, concluding that two oppositely charged fluids accounted for the attraction and

repulsions of electrified bodies. His ideas held sway for more than a century but have now been discarded in favor of your original idea—a single electric fluid, the *electrons*.

Electrons are extremely tiny—almost infinitesimally small—particles that surround the center of atoms (the ultimate indivisible unit of matter). Electrons whirl around each atom's center just as planets revolve around the sun. [Indeed, many today view atoms as miniature solar systems, an incorrect but understandable concept.]

Because electrons are located at the most exterior extremes of atoms, they are easily displaced from their position. Rubbing, for instance, dislodges electrons, as you did whenever you pressed wool against a rotating glass tube. Friction of our shoe against carpeted floors also rubs off electrons, causing a neutralizing spark to jump from our finger when we come near certain objects.

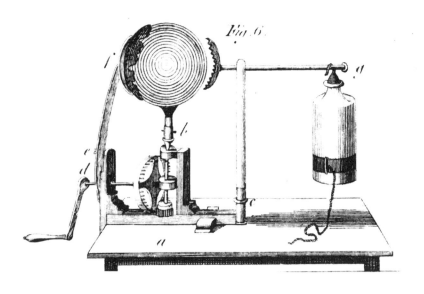

Rubbing a glass sphere removes electrons from the surface. The Leyden Jar (right) collects and stores the electrons

In a future email, I'll explain how we now generate vast quantities of flowing electrons, how we make them do mechanical work, light our homes and permit instantaneous communications across the planet, thanks in part to your research.

From: "Stuart Green" <stuartgreenmd@yahoo.com>
To: "Benjamin Franklin" <dr_benjamin_franklin@yahoo.com>
Subject: **Electricity in palsies**

Dear Doctor Franklin:

In some ways, Sir, you're lucky you lived in the 18th century because certain things you did would today put you in jail for practicing medicine without a license. I'm referring specifically to your experiments with *electrical stimulation*.

Do you recollect telling Dr. John Pringle how you got involved in the project? Here's what you wrote: "Some years since, when the newspapers made mention of great cures performed in Italy or Germany by means of electricity, a number of paralytics were brought to me from different parts of Pennsylvania and the neighbouring provinces, to be electrised, which I did for them, at their request."

John Pringle

We know that you sent an electrical apparatus to Jonathan Belcher, when he was the Royal Governor of New Jersey. Belcher wrote you that he was "sorry to inform you that when I came to open it the glass globe was broken all to pieces I suppose by the rough conveyance of it (in a wagon) from Burlington hither."

Luckily, the Reverend Arron Burr, Sr., a co-founder with Belcher of Princeton University (and father of America's third vice-president), had done some experiments with electricity and "made some use of the rest of the apparatus and with Mr. Burr's assistance have been electrified several times but a present with no alteration in my nervous disorder."

Likewise, you personally provided such treatment for James Logan, secretary to Pennsylvania's Royal Governor. He, like Belcher, had become partially paralyzed, suffering from what we now call a stroke, but was called in your time apoplexy. I presume you treated Logan in the manner you described to Pringle: Using Leyden jars to build up the charge, you "sent the united shock of these thro' the affected limb or limbs, repeating the stroke commonly three times each day."

James Logan

Your description of the initial response impressed me: "The first thing observed was an immediate greater sensible warmth in the lame limbs that had received the stroke than in the others…"

100

Your early results were encouraging. "The limbs too were found more capable of voluntary motion, and seemed to receive strength; a man, for instance, who could not, the first day, lift the lame hand from off his knee, would the next day raise it four or five inches, the third day higher, and on the fifth day was able, but with a feeble languid motion, to take off his hat."

However, the benefit didn't last; you informed Pringle: "These appearances gave great spirits to the patients, and made them hope a perfect cure; but I do not remember that I ever saw any amendment after the fifth day: which the patients perceiving, and finding the shocks pretty severe, they became discouraged, went home and in a short time relapsed; so that I never knew any advantage from electricity in palsies that was permanent." (Indeed, Dr. Franklin, from your time to the present, *permanent* palsies cannot be cured with electric stimulation.)

You ended by making a hopeful prediction about the future of electric shock therapy: "Perhaps some permanent advantage might have been obtained, if the electric shocks had been accompanied with proper medicine and regimen, under the direction of a skilful physician. It may be, too, that a few great strokes, as given in my method, may not be so proper as many small ones..."

As usual, you correctly surmised "permanent advantage" with many small shocks.

You'll be surprised to learn that one of your electricity proposals has proven valuable to patients suffering from melancholia (now called depression), although few today realize that you discussed *electroshock therapy* for that purpose in 1785. Your friend, the Viennese court physician Jan Ingenhousz, first suggested the treatment. You told him how you accidentally electrocuted yourself during an experiment: "I placed myself inadvertently under an iron hook which hung from the ceiling down to within two inches of my head. I neither saw the flash, heard the report, nor felt the stroke. When my senses returned, I found myself on the floor. I got up not knowing how that happened...I do not remember any other effect good or bad."

Ingenhousz told you of his similar experience with a jolt to the head, although with a surprisingly different outcome: "I lost all my senses, memory, understanding, and even sound judgment...I found I had entirely forgotten the art of writing and reading...This struck me with terror as I feared I should remain forever an idiot."

Rather than becoming an idiot, however, Ingenhousz did the opposite: "When I awoke the next morning...my mental faculties were at that time not only returned, but I felt a most lively joy in finding, as I thought at the time, my judgment infinitely more acute...I found moreover a liveliness in my whole frame, which I never had observed before."

Wondering if "such a commotion could be serviceable in mad people" induced Ingenhousz "to advise some of the London mad-doctors, as Dr. Brook, to try a similar experiment on mad men, thinking that...it might perhaps be a remedy to restore the mental faculties when lost; but I could never persuade anyone to."

You told Ingenhousz in 1785 that you "communicated that part of your letter to an operator encouraged by Government here to electrify epileptics and other poor patients, and advised his trying the practice on mad people..."

Over the next few decades, doctors tried electric jolts to the brain to treat certain mental diseases, and in some instances had success.

In the late 1930s, Italian physician Ugo Cerletti, after conducting experiments in a pig slaughterhouse to determine safety, began treating mental patients by administering electric shocks to the brain. It worked just as you and Ingenhousz proposed, with improvement noted immediately after the mind recovered from the initial effect of the jolt.

The technique became increasingly popular among today's *psychiatrists* (mad-doctors) until fairly recently, when new medications proved even more effective—and safer.

Today, an entire field of medicine—*clinical electrotherapy*—employs your electric fluid for numerous useful purposes. For example, a small electric machine doctors *implant in the body* treats heart rhythm problems. It delivers a light shock to the heart once a second to maintain a proper pulse. Such machines evolved from a remarkable discovery during your lifetime. I'll tell you about it in my next email.

Early 19th-century nerve stimulation machine

From: "Stuart Green" <stuartgreenmd@yahoo.com>
To: "Benjamin Franklin" <dr_benjamin_franklin@yahoo.com>
Subject: **M. Volta's experiment**

Dear Doctor Franklin:

Many great innovations result from fortuitous accidents, but none as strange as the twitching frogs-legs of Dr. Galvani. It happened in Italy in 1786, the year before the U.S. Constitutional Convention. During a lightning storm, Luigi Galvani (a physician and researcher) hung fresh frogs-legs on a balcony's railing. Galvani had already performed experiments at the University of Bologna on the electric stimulation of frog muscle, a well-known electrical curiosity. Perhaps he was preparing the frogs-legs for more experiments, or maybe he had a craving for *Grenouille Provençal*.

Galvani was startled to see the frogs-legs twitch occasionally, even though he hadn't applied electricity. He eventually realized that the muscles twitched whenever the frogs-legs, hanging from copper hooks, were blown by the wind against an iron railing.

Galvani conducted experiments and learned much about twitching muscles. Externally applied electricity wasn't necessary. Convinced that he had discovered something new, Galvani coined the term "animal electricity" to distinguish the effect from the "artificial electricity" generated by rubbing a rotating glass tube. (He called lightning "natural electricity.")

Prior to Galvani's fortuitous discovery, nobody could explain how the nervous system controlled muscle movement. The French philosopher and mathematician Renee Descartes, for example, assumed that nerves were water pipes or channels for fluid transport. Galvani realized that electricity provided the motive force.

Alessandro Volta, a University of Pavia professor and personal friend of Galvani, questioned the existence of animal electricity. Volta puzzled over the fact that metal probes had to be of *different* composition for a muscle to jump. He proposed that the dissimilar metals were the most important feature of the twitching demonstration. His idea put him on a collision course with Galvani—a rivalry stimulating far more than twitching muscles.

Volta was already an established electricity researcher when he confronted Galvani. In 1775, he invented the "electrophore," a charge-accumulating machine—forerunner of the modern capacitor—that replaced the Leyden jar for storage of a static electric charge. He later invented other static electricity machines.

Do you remember Jan Ingenhousz sending you a description of Volta's electrophore? Unlike your contemporaries, you realized that the machine was just another device to generate electric charge: "I thank you for the account you give me of M. Volta's experiment. You judge rightly in supposing that I have not much time at present to consider philosophical matters: But as far as I understand it from your description, it is only another form of the Leiden phial, and explicable by the same principles."

In 1791, the year after your alleged demise, The Royal Society of London awarded the Copley medal to Volta for his static electricity inventions, the same honor you received 38 years earlier. Volta's greatest accomplishment, however, occurred *after* he won the Copley.

Allesandro Volta

In 1794, Volta began a series of experiments that changed the flow of science. He immersed two different metals in a conducting solution and found that they would generate a steady flow of electricity without any animal tissue between them and without rotating glass either. He refined his invention by alternating discs of copper and zinc, each separated by saline-moistened cardboard. By connecting a wire to the top and bottom discs, Volta could create a substantial electric shock, enough to knock someone off his feet. In 1800, he first publicly demonstrated his "Voltaic pile" (today called an electric *battery*, borrowing your word for Leyden jars in series).

The flow of electricity from a Voltaic pile or battery doesn't last forever; the nature of the metals determines its strength and longevity.

Although eventually made a Count, posterity has given Alessandro Volta a kind of nobility far greater than that conferred by kings. Today, when people ask storekeepers for nine-*volt* batteries or inquire about the *voltage* of machines, they pay homage the Italian scientist who dipped dissimilar metal strips into salt water and generated a steady—but not perpetual—flow of electricity.

Luigi Galvani

We haven't forgotten Galvani either. Physiologists call continuous electric current applied to muscle *galvanic stimulation*. Because of the sudden way a muscle jumps when electrified, a person can be *galvanized* into action when stimulated by external events (such as unfair taxes). Time has shown that both Volta and Galvani were correct in their assertions: dissimilar metals *do* generate electricity without animal tissue when they are dipped in a conducting solution; and animals *do* create electric currents, although too weakly to measure in Galvani's time.

You'll be particularly proud of the way electricity helps today's paralytic patients. In fact, your single electric fluid plays an important role in modern health care in two broad categories: *electrodiagnostics* and *electrotherapeutics*.

As Galvani surmised, we internally generate low voltage electricity that travels along nerves to control many bodily functions. Modern equipment easily measures these currents, displaying them either numerically or graphically. Healthy and diseased hearts, for example, have characteristic patterns of electric currents that aid diagnosis and treatment.

Dr. Paul Zoll, a pioneering heart specialist from Boston, developed the first practical external cardiac muscle stimulator to treat heart stoppage. Some accused Zoll of interfering with the Almighty's plan by bringing the dead back to life. In many respects, the abuse he received mirrored what happened to you when certain clergymen claimed that your lightning rod thwarted the Lord's punishment. Fortunately for Zoll, a religious newspaper supported his effort by saying, "…God worked in many strange ways and this was one way of expressing the Divine Will." (You elegantly offered the same defense of your lightning rods and smallpox inoculation campaign.)

As with electric heart stimulation, your desire to help paralyzed patients with electric shocks is part of *electrotherapeutics*—using electricity in treatment. Considerable controversy has surrounded claims promoting one or another kind of electrotherapy. Physicians have, at times, used electric stimulation to treat every imaginable ailment.

At the turn of the 19th century, many doctors owned an office generator to administer therapeutic electricity. One device, for example, consisted of a chair insulated from the ground. The patient sat on the chair and was electrified. The practitioner used a metal rod to make static electric sparks fly out from the patient's body, assuring a cure. ('Twas unabashed quackery!)

Doctors in my specialty, faced with a broken bone that is not solidifying properly, may prescribe electrotherapy to stimulate healing.

Do you recall your patients described "a pricking sensation" after electric treatment? This phenomenon is now used for pain control; the pricking takes a patient's mind off the real source of discomfort.

When you administered three shocks a day to paralyzed people you were obviously hoping to stimulate recovery—still the goal of every electrotherapist. Advances in paralyzed muscle stimulation may soon fulfill your hope that "perhaps some permanent advantage might have been obtained, if the electric shocks had been accompanied with proper medicine and regimen, under the direction of a skilful physician."

Galvani's frog experiments

From: "Stuart Green" <stuartgreenmd@yahoo.com>
To: "Benjamin Franklin" <dr_benjamin_franklin@yahoo.com>
Subject: **Electricity and magnetism**

Dear Doctor Franklin:

During the two hundred years since your *mort-faux*, electricity has become the most common form of energy used by mankind. What started out as a parlor curiosity during your lifetime has turned into an essential component of our existence.

For thousands of years, the fire of rendered animal fat lit the world at night. Today, electricity serves that function. We use it to tell time, cook our food, help get us from place to place, and most significantly, communicate with each other. Even the most primitive people on the planet now use electricity in their daily lives. Because, as you know, electricity travels along metal wires, we today generate the electric charge in one location and transport it by wires to distant places for conversion into light, heat, or motion.

Han Christian Ørsted

An especially important discovery transforming electricity into a useful force occurred in 1820 when a Danish electricity experimenter named Hans Christian Ørsted made a startling observation. He passed electric *current* through a metal wire and noticed that a nearby compass needle rotated around and pointed directly at that wire. When Ørsted stopped the current, the needle spun back to the Earth's north-south line.

Ørsted recognized the importance of his observation. Recall how natural philosophers during your time (and earlier) attempted to contrast magnetism and electricity. In 1763, for instance, you said; "Professor Aepinus of Petersburgh [Russia], has published a work on magnetism and electricity, in which he endeavours to apply my theory of the latter to the explanation of certain phenomena in the former."

Shortly after Ørsted made his discovery that electricity affected magnetism, scientists worked out the relationship between the two forces. For a current to flow, Dr. Franklin, there must be pressure behind it, just as water needs a head of pressure to move in any particular direction. The pressure behind electric current flow is, as I've said, measured in units named after Volta; we call the pressure *voltage* and the measured units, *volts*.

(I'm sorry to inform you that we've no units in electricity called *franklins*. Now that you've revivified, you should petition the scholarly associations on your own behalf.)

The kind of electricity you studied we now label *electrostatics* because it deals with electrified bodies holding static charges. I realize that you might object, claiming that a charge can jump from place to place or be conducted by wires or silk threads. However, because of electricity's remarkable properties when it does move, we now refer to the branch of knowledge dealing with moving electric charges as *electrodynamics*.

During the 19th century, an English scientist did more than anyone else to harness electricity's motive power. Recall that Ørsted observed how a compass needle—or any magnet for that matter—rotated until it pointed directly at a wire through which electric current flowed. Turning off the current allowed the magnet to reorient itself to the Earth's magnetic field.

Michael Faraday, born in a village near London the year after you died, employed this principle to create an *electric motor*. By placing a second wire parallel to the first and conducting an electric current through the two wires in an alternating manner, Faraday found he could make a magnet rotate around a spindle.

He next affixed the magnet to the spindle set the construct in roller bearings. In this manner, the rotating magnet turned the spindle—now an axle—thereby converting electricity into rotary motion. Eventually, any driveshaft could be turned by electricity. Moreover, the motion remained smooth and reliable—as long as current flowed through the wires.

I realize that you also proposed an electric motor (which you called an "electrical jack") based on the attraction and repulsion of electrostatic charges. Although yours wasn't the first, a number of people have followed your directions and constructed your motor. They took a spindle-mounted wooden wheel and attached glass tubes around the wheel's periphery. By placing a small metal thimble at the end of each tube, these experimenters made points for attraction and repulsion as each thimble rotated towards—and then away from—two Leyden jars of opposite polarity near the protruding tubes.

Your electric wheel spins for half an hour after receiving an initial push. Such motors, however, aren't powerful enough for practical applications.

In 1831, Faraday went one step further. He discovered that, just as an electric flow induced magnetism, a moving magnetic field generated an electric current in a nearby wire. Faraday employed this principle to generate electricity.

Michael Faraday

Electric power generation is today a major industry. A modern version of the waterwheel turns a generator's axle. To assure a steady supply of moving water during dry seasons, nations have built great dams across rivers, creating artificial lakes to deliver flowing water year round.

We also use pressurized steam—created by boiling water with coal or petroleum oil—to produce electricity by spinning finned drive shafts connected to Faraday generators.

In certain countries with volcanic activity, citizens use the Earth's heat to convert water to steam, turning their generators. Windmills still dot the landscape in some locations. On our modern version, thin metal blades rotate the axle of an electric generator.

Once generated, electricity traverses thick metal wires (insulated from the ground on great towers) across the land to factories and homes and all manner of enterprises. There, electric motors convert the flowing electrons back into mechanical power—to spin wool, power looms, stir chemicals and lift heavy objects. Likewise, some of the electricity passes through thin wires, causing them to glow luminously, providing us with bright and dependable lighting. In other locations, thicker wires made of resistant metals serve as cooking elements in electric stoves and ovens.

Here's another useful feature of magnetism's induction of electricity: coiling the wire enhances the effect by acting like a bar magnet. Engineers boost the magnet effect by putting a steel rod the coil's center, creating an *electromagnet*, an exceptional device. The imposed magnetism lasts only as long as current flows through the coil.

Do you recollect discussing this temporary effect of magnetism on steel compared to its permanent influence on iron? In 1763 you said that Aepinus "supposes magnetism to be a particular fluid equally diffused in all iron, easily flowing and easily moving in soft iron, so as to maintain or recover an equilibrium. But in hard iron, *i.e.*, steel, moving with more difficulty."

The electromagnet permits fabrication of an electrically-controlled switching mechanism. Imagine a flat plate of steel, resting on a pivot at one end and suspended from a spring on the other. Now envision an electromagnet on the opposite side of the plate from the spring. Electrifying the magnet pulls the plate and holds it fast to the magnet until the current stops flowing. With interruption of current, the spring pulls the plate back to its resting position.

By applying an interrupting current to such a device, the plate click-clicks towards and away from the electromagnet. Indeed, one can vary the length of time the circuits closed, thereby creating short and long click sounds.

Samuel F. B. Morse, born the year after you died, played a major role in the history of communications by developing (in 1832) a clicking electromagnet to send messages over wires. The circuit is opened and closed in one location and a metal bar somewhere else does the clicking.

Morse also devised a code of clicks to enable easy communication. It's amazing how quickly operators of his system learned the pattern; they transmitted messages across distances of hundreds—or even thousands—of miles using Morse's Code.

After several failed tries, an insulated wire was laid along the Atlantic Ocean's floor in 1866. Thus did electricity end America's isolation from Europe and the rest of the world.

Some other remarkable applications of electricity had their seeds planted during your own era and by one of your closest friends, the Reverend Joseph Priestley. I'll describe his valuable contribution in my next email.

From: "Stuart Green" <stuartgreenmd@yahoo.com>
To: "Benjamin Franklin" <dr_benjamin_franklin@yahoo.com>
Subject: **The conducting quality of some kinds of charcoal**

Dear Doctor Franklin:

Do you remember discussing charcoal with Joseph Priestley? History credits him with discovering the conductive property of that substance, which has proven useful in many applications.

If two sticks of charcoal in an electric circuit touch each other, electricity will flow through one to the other as though they were wire conductors. By slowly pulling the sticks apart, current will jump the gap, creating a continuous bright light (called an *arc*), an early type of electric illumination. Although glowing wires and other technologies have since replaced charcoal arc lighting, it's still used whenever intense white luminosity is needed.

Carbon also has an unusual property not found with most substances that conduct electricity: the resistance to electric flow varies with pressure applied to its surface. This principle serves us in the carbon *microphone*, a device that converts the pressure variations of sound waves into comparable fluctuations in electric resistance.

A Scottish-born teacher of deaf students, Alexander Graham Bell, created the first practical instrument for transmitting speech and other sounds along electric wires. His first *telephone* employed a flexible metal diaphragm attached to an electromagnet, which connected to electric batteries. He completed the circuit with an identical diaphragm-electromagnet instrument elsewhere. When Bell transmitted "Mr. Watson, come here, I want to see you" to his assistant on March 10, 1876, he made history.

Unfortunately, Bell's metallic diaphragm didn't convert sound to resistance very well. It took another inventor—perhaps the greatest who ever lived—to propose carbon for the microphone. Thomas Edison, a New Jersey man, in 1886 invented a microphone employing charcoal granules that greatly improved the quality of sound transmission over wire.

People today remember Edison as the inventor of the first commercially successful electric *light bulb*, an illuminating application of the glowing-wire principle using filaments of carbonized fabric. Many of his inventions, including machines that permanently record sound and moving images, have had great impact on the human experience.

More about electricity in my next email, but for now, I've got to go to work. (I'm late!)

Telephone patent application

A. G. Bell

Thomas A. Edison

Light bulb patent application

From: "Stuart Green" <stuartgreenmd@yahoo.com>
To: "Benjamin Franklin" <dr_benjamin_franklin@yahoo.com>
Subject: **Some principle yet unknown to us**

Dear Doctor Franklin:

What I respect most about you, Sir, is your open-minded approach to questions of science. When possible, you devised simple yet elegant experiments to either confirm or refute your conclusions. If your research failed to support a conjecture, you willingly abandoned it without feeling compelled to come up with an alternate hypothesis. A perfect example of this attitude involves your proposition that the oceans generated lightning's electricity.

As you explained in a 1752 letter to Bostonian James Bowdoin, you were impressed by the glow of light that sometimes appears when oars and paddles stir up seawater at night. You wrote, "my supposition that the sea might possibly be the grand source of lightning, arose from the common observation of its luminous appearance in the night on the least motion; an appearance never observed in fresh water."

Knowing that rubbing crystals produces an electric charge, you proposed that friction between the water and salt globules generated electricity. That friction, according to your assumption, created a "subtile electric fluid," which "electrified surface of the sea, " and subsequently the clouds.

However, you told Bowdoin that you "endeavored in vain to produce that luminous appearance from a mixture of salt and water agitated; and observed that even the sea water will not produce it after some hours standing in a bottle." Because you could not create the glow by mixing salt and water, you willing admitted that the light might "proceed from some principle yet unknown to us (which I would gladly make some experiments to discover, if I lived near the sea) and I grow more doubtful of my former supposition"

In reply, Bowdoin hit on the correct explanation: The luminosity "might be caused by a great number of little animals floating on the surface of the sea, which on being disturbed might...exhibit a luminous appearance in manner of the glow-worm or fire-fly..."

Today, Dr. Franklin, we call the glow created by living creatures *bioluminescence*, and a similar light emitted when certain chemicals mix, *chemoluminescence*. In both cases, light appears unaccompanied by heat, an unusual situation, I'm sure you'll agree.

This phenomenon also perplexed you friend Ezra Stiles, later a president of Yale College. Recall that he wrote you about it in 1757, saying, "The philosophy of light and fire, heat and cold has hitherto been a mystery to me." Questioning the ancient notion that all matter was composed of the four elements—earth, air, fire and water—he wrote, "I am half persuaded that light is an element different from fire."

Stiles seemed particularly interested in the observation that a glow-worm's light was "luminous, but not hot; the flame of a candle is however luminous and hot." He nevertheless recognized the obvious association between heat and color, as with fire. Moreover, Stiles commented to you about the relationship between the quantity of heat and the color of the light emitted: "When the action of the fire is less vehement," he said of heated iron, "the bodies exhibit other colors—generally red," whereas the hottest pieces "whence they in this state appear white, or a luminous glow."

We've since learned much about heat and light; they are indeed related.

I'll tell you more about it in my next email.

James Bowdoin

From: "Stuart Green" <stuartgreenmd@yahoo.com >
To: "Benjamin Franklin" <dr_benjamin_franklin@yahoo.com>
Subject: **As thro' a vacuum**

Dear Doctor Franklin:

Today's email covers a complex topic—heat and light—but I'll do my best to explain it. Newton, you'll recall, determined that light moved through space in waves, resembling those on the sea's surface. (Ezra Stiles referred to Newton's light waves as "The undulatory vibrations of the luminous fluid.") As with water and sound waves, light waves vary in frequency, height and wavelength

Now we come the heart of the question: How is light generated when a heated object glows? I mentioned in a prior email that your electric fluid consists of *electrons*—particles that reside in shell-like *energy levels* surrounding each atom's center (called the *nucleus*). Each shell has the capacity for a limited number of electrons, with the shell closest to the atom's center holding the fewest.

When heated, the outer electrons of atoms jump out to a higher energy level; that is, to an empty shell beyond the ones surrounding the nucleus. Electrons are unstable in that position, so each one immediately drops back to its former shell level. Just as water, when pumped up to a rooftop cistern, yields energy as it flows down, so do electrons give off energy when resuming their former position.

As the heating continues, electrons jump out and drop back repeatedly. This out-and-in oscillation releases energy in the form of *electromagnetic waves*, either as visible light, radiant heat, or many other kinds of invisible radiation. All such waves exist along an *electromagnetic spectrum*, a greatly expanded version of Newton's prismatic light spectrum.

The wavelength of light is extremely short: 50 million peaks and troughs to the inch. Some electromagnetic waves, however, are surprisingly long—up to 300 feet between peaks. These *radio* waves can travel through air to great distances, allowing *wireless* transmission of energy. By modulating either the amplitude or frequency of such waves, we've learned how to communicate through the atmosphere to places far away.

Other forms of radiant energy have wavelengths a thousand times shorter than visible light, allowing the waves to penetrate solid matter. We use such *X-rays*—discovered by a German researcher named Wilhelm Roentgen in 1895—to see through flesh to the bones and internal organs.

Electrons, like light, radio, and X-ray waves, can be also produced in a stream that travels in a straight line. In fact, both lightning and the spark of your apparatus light up when flowing electrons strike gases in the air blocking their pathway.

Particles of air thus inhibit the movement of electrons, Dr. Franklin, a restraint eliminated by generating electrons in a vacuum. You clearly knew this principle when you wrote, "An electric atmosphere cannot be communicated at so great a distance thro' intervening air, by far, as thro' a vacuum." Without air particles to impede them, electrons flow freely across a void from a donator of electrons—called the *cathode*—to a receiver, the *anode*.

Just as a glass lens bends light rays, a magnet can curve a stream of electrons. The electron microscope, mentioned in an earlier email, uses electromagnets to focus and then refract an electron beam in a vacuum tube, creating an enlarged image of a specimen placed in the beam's path.

A similar device paints the screen image on certain computers. Recall that Francis Hauksbee's vacuum tube (containing a small amount of mercury) glowed when electrified. Certain computer screens are also vacuum tubes, coated on the inside with chemicals that glow when struck by electrons. When energized, a cathode within the tube generates a beam of electrons that stream towards the screen. Electromagnets control the beam's *position*, which sweeps across the screen in a line-by-line pattern, 25 times per second. The electron beam's *intensity* varies with the energy at the cathode.

Looking closely at such a screen reveals the individual dots that glow with luminosity governed by the beam's strength at the moment it strikes each dot. A pure white color forms when the chemical glows maximally, and black is the absence of any glow. Different combinations of adjacent white and black dots create shades of grey.

Such computer screens generate *colors* by using chemicals that radiate specific hues when struck by electrons.

Those same dots can be created with ink on a page, although in that situation, black is the presence of ink and white occurs when the underlying page comes through, just as with your own press. Lead type, the hallmark of printing since Gutenberg, is gradually being eliminated by modern technology. The computerized printing machine shoots extremely small ink globs at a piece of paper, literally painting letters on the page.

All of this computer technology has occurred within the past 50 years. Had you been revived a half century earlier, Dr. Franklin, you'd have found a more familiar environment, compared to what you see today.

Speaking of computer images, in 1951 historians at Yale University started to assemble and publish all documents written by you or to you—a multivolume effort that continues today. My friend Ellen Cohn presently heads the project.

Not long ago another scholarly group, The Packard Humanities Institute, created an electronic compilation of the Yale undertaking for use on a computer. They kindly supplied me with a pre-publication copy.

But for *The Papers of Benjamin Franklin* and the computer-friendly version sent by PHI, I couldn't proceed with these emails.

CHEMISTRY

✉ You have the philosopher's-stone   119

✉ Wind generated by fermentation   123

✉ Monsieur Geoffroy of the Royal
   Academy of Sciences   125

✉ This air is not fit for breathing   129

✉ There is nothing unhealthy in the
   air of woods   131

✉ A candle burned in this air with a
   remarkably vigorous flame   135

✉ You have set all the philosophers of
   Europe at work upon fixed air   137

✉ Murdered in mephitic air so many
   honest harmless mice   139

✉ Method of making salt-petre   141

✉ The power of man over matter   145

✉ On The bad effects of lead   149

18th-century chemistry laboratory

From: "Stuart Green" <stuartgreenmd@yahoo.com >
To: "Benjamin Franklin" <dr_benjamin_franklin@yahoo.com>
Subject: **You have the philosopher's-stone**

Dear Doctor Franklin:

You lived during a remarkable epoch in the history of chemistry. Your birth in 1706 occurred while alchemists were still concocting magic potions to gain immortality. By the time of your encaskment in 1790, a "new chemistry" had evolved, complete with modern-looking equations, a reasonably correct theory of fire and the rusting of metals, the insight that our food undergoes slow combustion in our bodies, and the concept that respiration both supplies and eliminates gases involved in life's processes.

Moreover, many of the researchers who made the momentous discoveries of 18th-century chemistry and physiology were either your casual acquaintances, your close friends, or, in the matter of food's internal combustion, yourself.

The relationship between *physiology* (the study of bodily function) and *chemistry* (the present spelling of chymistry) during your era, Dr. Franklin, was especially close in the field of respiratory science. The remarkable advances in chemistry that occurred during your lifetime often required a living animal—bee, mouse, guinea pig—to determine if a newly discovered gas maintained life or caused the creature's demise. A burning candle served a similar purpose, but proved no substitute for living creatures because certain unidentified gases supported combustion yet couldn't sustain life.

Thus, without live-animal experiments, we'd still lack the most basic knowledge about chemical reactions and why they happen.

During the five hundred years we call the Dark Ages, European and especially Arab metalworking artisans learned how to make and identify acids, alkalis and the salts of metals. Their primary goal was to extract metals from ores more economically. They discovered that certain metals—gold, silver, copper, lead, tin, iron and mercury—once purified seemed as elemental stuff, not capable of further reduction to the four Aristotelian substances (earth, air, fire and water) that presumably formed all matter. This, in turn, challenged established doctrine about Greek-derived concepts.

But, just as pre-Copernican astronomy became more complicated before it could be simplified, the beauty and symmetry of the Greek four-element system got worse before it got better.

In the early 1500s, Paracelsus (a bombastic but influential self-trained physician), and others overthrew the four-element concept by adding three "principles"—sulfur, mercury, and salt—which could influence the original four. Thus, air could become sulfurous, resulting in the foul smelling "air" we associate with rotten eggs. Water, upon becoming sulfurous (now *sulfuric acid*), burned the flesh and dissolved metals.

By the mid 1500s, insightful chemistry workers and natural philosophers objected to air's status as an indivisible element. A lit candle placed under an inverted glass bowl would burn for a while and then extinguish, yet the bowl still appeared to have air in it, suggesting that something happened to change the bowl's air. Likewise, a mouse placed under a bowl behaved like a candle, doing well for a while but then becoming agitated, searching for a way out, and finally dying.

Thus, candles and creatures needed some component of air, which they consumed.

Although many people today assume that chemistry evolved from alchemy, historians contend that the two disciplines existed side by side during the centuries before your birth. The alchemists, as you know, sought the "philosopher's stone" that could turn a common metal—such as lead—into gold.

A liquid extract of philosopher's stone, the so-called Elixir of Life would have conferred either a long disease-free existence or prevented death altogether for someone who consumed it.

Alchemy reached its pinnacle of activity during the century before yours, with many individuals searching vainly for the philosopher's stone. (You obviously were skeptical about the claims of alchemists. I like the practical way *Poor Richard* admonished readers: "If you know how to spend less than you get, you have the Philosopher's-Stone.")

18th-century chemical (cement) manufacturing facility
From Humphreys' *Nature Display'd*

120

Chemists, on the other hand, dealt with more mundane goals—improving metallurgy or leather tannery or pigment making. By the late 17th and early 18th centuries, for example, chemists could create the "calx" of almost every known metal or mineral. (A calx, as you know, is the powdery substance obtained by roasting or burning something.)

Jan Baptista Van Helmont, as you may recall, was the 17th-century Belgian physician who launched the chemical revolution by challenging established dogma. He questioned the notion that fire was a basic element, preferring to consider it instead an "imponderable substance." In his view, fire shared with light and heat (and later electricity) the property of having no weight yet was somehow involved in chemical or physical reactions.

Jan Baptista Van Helmont

Van Helmont likewise determined that air contained fractions of differing properties, some of which were downright dangerous. He knew that the fumes bubbling off during beer fermentation could suffocate a person. He also concluded that the substance could extinguish a flame, and, moreover, was the same matter given off by a burning candle. He coined the term "gas" (from the Greek word for chaos) to describe this material, which he claimed was distinct from air, although mixed in with it. Equally remarkably, Van Helmont deduced that an infamous Italian cave called the *Grotto del Cane* contained this harmful gas.

Do you remember the *Grotto del Cane* near the volcano Mt. Vesuvius? It has a peculiar feature that certainly caught your fancy. If a person walked a dog into the cave, that individual experienced no symptoms but the dog, after a short period of time, became first sleepy and then unconscious. The animal, if not removed promptly from the cave, soon died.

Van Helmont reasoned that a heavy gas produced by Vesuvius settled to the cave's floor to a depth that would suffocate a dog but not a standing person.

You obviously knew about Van Helmont's ideas. Your March 21, 1734, *Pennsylvania Gazette* advertised Van Helmont's *Works* for sale, along with a few dozen other classic tomes. Moreover, you must have remarked about the *Grotto del Cane* to some of your friends because at least one of them erroneously believed that *you* first figured out how the *Grotto del Cane* killed dogs.

Your friend Alexander Small, the Scottish surgeon, wrote *On Franklin's Views of Ventilation*, which must have given you considerable satisfaction. To remind you, Small started out by complaining, "I do not know that we have, in any author, particular and separate directions concerning the ventilating of hospitals, crowded rooms, or dwelling-houses." Small next claimed that, "the want of such general information, on these subjects, has induced me to endeavour to recollect all I can of the many instructive conversations I have had upon these matters with that judicious and most accurate observer of nature, Dr. Benjamin Franklin."

After mentioning that when people are confined to a small room for a while, the air within no longer sustains life, Small said, "Dr. Franklin was, if I mistake not, the first who observed, that respiration communicated to the air a quality resembling the mephitic [noxious]; such as that of the *Grotto del Cane* near Naples. The air impressed with this quality rises only to a certain height, beyond which it gradually loses it."

Small claimed that you even made a demonstration showing how the component of air that snuffed out a candle was heavier than the component that supported combustion.

More about this in a future email.

From: "Stuart Green" <stuartgreenmd@yahoo.com>
To: "Benjamin Franklin" <dr_benjamin_franklin@yahoo.com>
Subject: **Wind generated by fermentation**

Dear Doctor Franklin:

I know you admired Robert Boyle and the quality of his research in numerous areas of physics and chemistry; he wrote dozens of books, most of them classics in science. With his assistant Robert Hooke (author of *Micrographia*), Boyle improved the vacuum pump, an especially useful tool for research involving gases. He also established what we now call Boyle's Law which defines the relationship between the pressure of a gas and its volume.

Robert Boyle

Boyle's 1661 book *The Sceptical Chymist* redefined the notion of an "element" as a substance that cannot be further broken down by chemical reactions. He concluded that none of the four Greek elements fit such a criterion. Boyle went further: He insisted that experimentation was the path to understanding.

You also respected Boyle because he tried to find common ground between the spiritual and real worlds. I wish that some of today's dogmatic clerics and their followers would read your statement about Boyle's religious tenets and the relationship of science to faith: "tho' ignorance may in some be the mother of devotion, yet true learning and exalted piety are by no means inconsistent."

Middlesex clergyman Stephan Hales was another early chemist whose name appears repeatedly in your papers, even though it seems you two never met. He was, after all, an old man—from Newton's generation—when you arrived in London for the first time in 1724.

Hales' method of collecting gases by water displacement

Hales set the stage for subsequent research because he figured out how to easily collect gases in upside-down water-filled jars, an invaluable procedure for all who followed in the history of chemistry.

Hales set up his apparatus and heated plants until they roasted, collecting the resulting gas in the process. He never determined the nature of the gas, only that it seemed somehow "fixed" in plants, so he called the gas "fixed air" (now *carbon dioxide*). He also determined that growing plants absorbed something from the air that they "fixed" into themselves, which reduced the air's volume

Hales studied other features of plant and animal physiology, measuring both sap and blood pressure, the first person to do so. You obviously knew of Hales when you wrote Cadwallader Colden about perspiring and imbibing pores: "Dr. Hales helped me a little, when he informed me, (in his *Vegetable Statics*) that the body is not always in a perspirable but sometimes in an imbibing state…"

Colden seemed amazed that you were, in 1745, unfamiliar with the gas given off during fermentation: "It surprises me a little that wind generated by fermentation is new to you since it may be every day observed in fermenting liquor. You know with what force fermenting liquors will burst the vessels which contains them if the generated wind have not vent…That air is generated by fermentation I think you'll find fully proved in Dr. Hales's analysis…"

From: "Stuart Green" <stuartgreenmd@yahoo.com>
To: "Benjamin Franklin" <dr_benjamin_franklin@yahoo.com>
Subject: **Monsieur Geoffroy of the Royal Academy of Sciences**

Dear Doctor Franklin:

A substantial part of chemical experiments during the 17th and 18th centuries involved the search for medical preparations to relieve symptoms of illness. Paracelsus called this pursuit *iatrochymistry* (doctor-chemistry), the forerunner of modern pharmaceutical science. The work of physician-chemists began to yield significant results in Germany, France and England while you were still young.

The importance of alchemy declined after Prussian physician Georg Ernst Stahl (1660-1734) developed a "rational" chemistry. Stahl attacked alchemists for their mysticism, vanity, secretiveness, charlatanism, impiety and greed. He lauded chemistry, on the other hand, as a rational enterprise searching for answers to the betterment of mankind and the glory of God.

Chemists, Stahl correctly pointed out, sought improvement in medications, metallurgy, beer making, glass blowing and the like.

Stahl modified an earlier theory and gave the name "phlogiston" to the liberated component of any flammable material. To Stahl, phlogiston was a weightless imponderable substance that, although involved in combustion, differed from fire. Phlogiston can't be detected by the senses; fire can. The more phlogiston in a substance, the brighter it burns—or the longer the flame lasts.

Charcoal and inflammable oils thus contain lots of phlogiston. Things that don't burn either have no phlogiston or lost it during earlier combustion.

I assume, Sir, you saw value in the phlogiston theory because it explained seemingly unrelated phenomena—a good test for any hypothesis. For example, a lit candle placed in a closed bell jar would burn for a while and then go out, even though the candle had plenty of phlogiston left (because it burned if reignited outside the jar). Stahl explained: when the jar's air became saturated with phlogiston, it took up no more, extinguishing the candle.

Stahl's phlogiston also accounted for certain observations in respiratory science: a mouse placed in a bell jar would live for a while and then die. According to phlogiston theory, the enclosed air gradually became saturated with phlogiston from the rodent's exhaled breath until it could absorb no more, leading to death.

Finally, phlogiston theory explained the rusting of iron, which gives up phlogiston in the process, leaving a calx (rust). Here, phlogiston encountered its fatal flaw: careful weighing revealed that iron, after rusting, became *heav-*

ier. Don't worry, explained Stahl's followers; they concluded that phlogiston was lighter than air, and so it buoyed up materials until it was released. This concept, although clever enough, didn't fit with the notion that phlogiston was imponderable, and thus not measurable, even as negative weight.

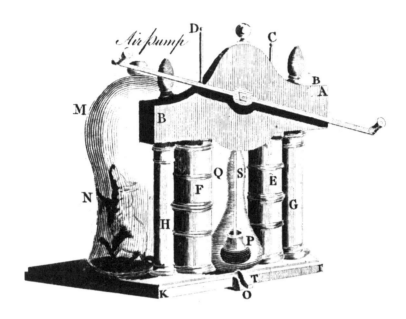

Air pump withdrawing air from bell jar containing mouse

Nevertheless, phlogiston theory lasted nearly one hundred years, Dr. Franklin, as the most widely accepted explanation for the peculiar features of metal calcinations, the combustion of materials, and the physiology of plants and animals. Stahl died the year after you began publishing *Poor Richard's Almanack*, no doubt thinking he had solved one of nature's most challenging puzzles.

Around the time Stahl established his phlogiston theory, Etienne-Francois Geoffroy, a French professor of medicine and chemistry, put together *Table of Affinities* (1718), an attempt to organize chemical reactions in a logical fashion. Geoffroy began a tradition of French chemistry that led directly to the experiments of your friend Antoine Laurent Lavoisier, the universally acknowledged father of modern chemistry.

You doubtless knew of Dr. Geoffroy because you referred to him in your 1756 *Alamanak* describing the treatment of bloody diarrhea thusly: "Monsieur Geoffroy...says, of all the preparations of glass of antimony this is doubtless the most perfect..."

Another significant development arising from iatrochemistry occurred when your acquaintance, the Scottish physician Joseph Black, in the early 1750s, analyzed the heartburn antacid Magnesia Alba by roasting it. He created a residue and gas that he correctly assumed was the same as Hales' discovery.

Unlike previous experimenters, however, Black, by weighing both the starting and resultant materials in his experiments, began *analytic chemistry*, essential to understanding how substances reacted with each other. Through his research, Black pointed the way for later scientists to extract gases from minerals and other substances.

Joseph Black

I know you met Joseph Black at least twice during trips to Scotland, the first in 1759 and again in 1771. Please tell us about those encounters when you get a chance. Did you two discuss chemical or electrical experiments, or did you merely exchange pleasantries? We've no letters between you and Black, which to me seems a bit odd. After all, you corresponded with many notable scientists, so why not Dr. Black?

Once Black demonstrated how to unlock gas from stones and metals, other chemists followed suit and began collecting whatever fumes emanated from burning or roasting materials. Prior to Black's research, such smoke went, quite literally, up the chimney. Likewise, future chemists weighed all the materials involved in chemical reactions. They gradually realized that, just as Newton's work revealed conservation of momentum and your discoveries established conservation of electric charge, Black's research found a conservation of material in chemical reactions—if the experimenter accounted for all products of the interaction.

The next group of scientists who distinguished themselves as chemists—Henry Cavendish, Joseph Priestley, Karl Scheele and Antoine Lavoisier—are rightfully called the *pneumatic chemists* because so much of their work involved gases. Moreover, three of them were among your acquaintances. If you've forgotten what they did, fear not, Dr. Franklin: I'll continue my efforts to refresh your memory of events and persons from your time *before* you'll have to face the public.

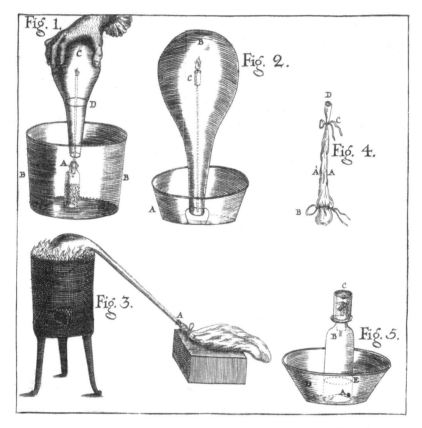

Karl Scheele's equipment for collecting gases and studying the respiration of bees (figure 5).

From: "Stuart Green" <stuartgreenmd@yahoo.com>
To: "Benjamin Franklin" <dr_benjamin_franklin@yahoo.com>
Subject: **This air is not fit for breathing**

Dear Doctor Franklin:

I'm curious to know if what I've read about Henry Cavendish's odd behavior was true. I understand that in spite of considerable inherited wealth he limited personal contact, especially with females. Did he really communicate with his housekeepers by leaving messages on the dinner table?

Certainly you formed an opinion of Cavendish while you two served on the Purfleet Commission. Regardless of his behavior, his eminence lives on: a famous British research laboratory honors his name.

In 1766, the same year you struggled in London to get the Stamp Act repealed, Henry Cavendish reported his research on gases. He described how he obtained "inflammable air" by dissolving iron in what we now call *hydrochloric acid*. As soon as he put a flame to the gaseous product of that reaction, it exploded. He at first assumed that his gas was pure phlogiston, but soon realized he was wrong: Cavendish had discovered a distinct gas that Lavoisier later called *hydro-gine* (meaning water maker).

After one explosion in a laboratory bottle, Cavendish noted water droplets inside the glass. He correctly guessed that his gas had combined with something in the atmosphere to form water. Cavendish thus discovered that water is formed from two gases.

Correct me if I'm wrong, Sir, but didn't Joseph Priestley become your close friend after you two met in London in the early 1760s?

Priestley, born of humble parents near Leeds, became a Nonconformist minister who tinkered with the electricity generators of his day and made some important discoveries. First, as I mentioned in an earlier email, he determined that carbon conducted electricity—the basis of today's carbon arc lighting and the carbon microphone. Second, he learned that salt conducted electricity but sugar didn't. And third, Priestley discovered that an electric charge collected on the outside of a hollow metal object. He was clearly pleased to make your acquaintance, calling you "the father of modern electricity" for your discoveries in the early 1750s.

Priestley's 1766 *The History and Present State of Electricity, with Original Experiments* earned him election to the Royal Society after you sponsored him. At around that time, he moved into a residence adjacent to a brewery. There, with abundant fixed air (*carbon dioxide*) from fermentation, Priestley started research on gases to the everlasting benefit of mankind. Priestley soon observed that both a candle and a mouse died quickly when placed in the gas.

When Priestley collected the brewer's gas by the inverted jar technique, he realized that the bubbles give water a rather pleasant taste. He also noticed that water thus produced shared taste and refreshing quality, with the naturally effervescent water obtained at Bad Pyrmont, a mineral spring in lower Saxony.

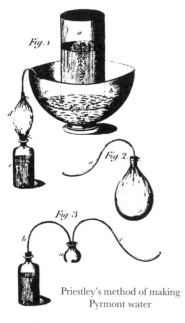

Priestley's method of making Pyrmont water

You certainly enjoyed the claimed benefits of Pyrmont water. In the early summer of 1766 you wrote Deborah, "To-morrow I set out with my friend Dr. Pringle (now Sir John) on a journey to Pyrmont, where he goes to drink the waters; but I hope more from the air and exercise, having been used, as you know, to make a journey once a year."

I've learned much about Priestley's research by reading your 1772 letter to John Hawkesworth appraising him of the discoveries: "This air is not fit for breathing; flame is extinguished by it; and, taken into the lungs, it instantly extinguishes animal life: But taken into the stomach is deemed salutary, as in Pyrmont Water which contains much of it."

Priestley published his observations about fixed air in a 1772 book *Impregnating Water with Fixed Air in order to communicate to it the particular Spirit and Virtue of Pyrmont Water*. In it, he suggested using an air pump to force even more brewer's gas into the water increased the effervescence and enhanced the flavor.

Two years thereafter you wrote to Dr. Benjamin Rush in Philadelphia, "This artificial Pyrmont Water, like the artificial magnet, is preferred to the natural, and for the same Reason, because it is cheaper, may be made stronger, and can be refreshed at any time with additional virtue."

Two hundred and fifty years later, members of the International Carbonated Beverage Bottlers Association reprinted Priestley's *Pyrmont Water* pamphlet, which they venerate as the Holy Document of their industry.

From: "Stuart Green" <stuartgreenmd@yahoo.com>
To: "Benjamin Franklin" <dr_benjamin_franklin@yahoo.com>
Subject: **There is nothing unhealthy in the air of woods**

Dear Doctor Franklin:

More today about your friend Joseph Priestley. Wanting to extend his research to the vegetable kingdom, Priestley placed a living sprig of mint in the brewery's gaseous output, fully expecting the plant to die in short order. Instead, to his surprise, the mint flourished. Although Priestley misinterpreted his discovery, he demonstrated the phenomenon to you in June 1772.

I suspect that you immediately recognized the importance of mint's reaction to fixed air because as soon as you left, Priestley performed more experiments. He excitedly wrote you about them a few weeks later: "I presume that by this time you are arrived in London, and I am willing to take the first opportunity of informing you, that I have never been so busy, or so successful in making experiments, as since I had the pleasure of seeing you at Leeds."

Following your visit, Priestley "put a mouse to the [brewer's] air in which it [the mint] was growing…seven days" and the mouse survived "five minutes without showing any sign of uneasiness, and was taken out quite strong and vigorous." However, another mouse died, "after being not two seconds" in the brewery's gas "without a plant in it." Priestley also mentioned that, "I have completely ascertained the [mint's] restoration of air in which, tallow or wax candles, spirit of wine, or brimstone matches have burned out by the same means."

Although neither you nor Priestley realized that plants use carbon dioxide and excrete *oxygen* to the atmosphere, you saw past mouse and mint to the significance of it all. Your reply to Priestley, if better known, would label you a farsighted sage even without your electricity discoveries: "That the vegetable creation should restore the air which is spoiled by the animal part of it, looks like a rational system," comparing such a gas cycle to the rain-evaporation-rain sequence and the foodstuff-manure-foodstuff succession.

You ended the letter with this statement that should add "*Father of Environmental Ecology*" to your accolades: "I hope this will give some check to the rage of destroying trees that grow near houses, which has accompanied our late improvements in gardening, from an opinion of their being unwholesome. I am certain, from long observation, that there is nothing unhealthy in the air of woods; for we Americans have everywhere our country habitations in the midst of woods, and no people on earth enjoy better health, or are more prolific."

Priestley's religious affiliation and politics created certain problems for him around this time, perhaps to the betterment of mankind. His church was Unitarian, Priestley having switched over to that belief system. He was briefly considered for a berth on Captain Cook's voyage of discovery but the government's ministers wouldn't send a man on an official mission who denied the Doctrine of the Holy Trinity.

Likewise, when you tried to get Priestley a teaching position at Harvard, it appears that the same thing happened.

Joseph Priestley

I'm particularly interested in your 1772 visit to Leeds when you told Priestley of an unusual phenomenon you heard about while passing through New Jersey eight years earlier. Priestley wanted to include the information in his forthcoming book *Experiments and Observation on Different Kinds of Airs*. You obliged.

Although it must have seemed incredible at the time, you learned "that by applying a lighted candle near the surface of some of their rivers, a sudden flame would catch and spread on the water, continuing to burn for near half a minute."

I assume you were skeptical about the presence of a flammable river since you wrote: "I could form no guess at the cause of such an effect, and rather doubted the truth of it."

EXPERIMENTS

AND

OBSERVATIONS

ON DIFFERENT KINDS OF

A I R,

AND OTHER BRANCHES OF

NATURAL PHILOSOPHY,

CONNECTED WITH THE SUBJECT.

IN THREE VOLUMES;

Being the former Six Volumes abridged and methodized, with many
Additions.

By JOSEPH PRIESTLEY, LL. D. F. R. S.

AC. IMP. PETROP. R. PARIS. HOLM. TAURIN. ITAL. HARLEM. AUREL.
MED. PARIS. CANTAB. AMERIC. ET PHILAD. SOCIUS.

V O L. I.

———

Fert animus caufas tantarum expromere rerum,
Immenfumque aperitur opus.
LUCAN.
Motto to the Firft of the Six Volumes.

———

BIRMINGHAM,

PRINTED BY THOMAS PEARSON;

AND SOLD BY J. JOHNSON, ST. PAUL'S CHURCH-YARD, LONDON.

M DCC XC.

You checked out the story with a friend who confirmed everything and even told you how he went to "choose a shallow place, where the bottom could be reached by a walking-stick, and was muddy; the mud was first be stirred with the stick, and when a number of small bubbles began to arise from it, the candle was applied. The flame was so sudden and so strong, that it catched his ruffle [collar] and spoiled it."

In hindsight, you erroneously speculated on the cause of the flames: "New-Jersey having many pine-trees in different parts of it, I then imagined that something like a volatile oil of turpentine might be mixed with the waters from a pine-swamp, but this supposition did not quite satisfy me." You concluded by suggesting that "The discoveries you have lately made of the manner in which inflammable air is in some cases produced, may throw light on this experiment, and explain its succeeding in some cases, and not in others."

We now know that swamp gas (*methane*) is not the same as Cavendish's inflammable air (hydrogen).

Priestley's scientific work came to the attention of Lord Shelburne, a wealthy nobleman and later, the prime minister who agreed to your peace terms ending the Revolutionary War. The year after Priestley demonstrated to you his mint sprig experiment, Shelburne asked him to move into his baronial estate as a full-time literary companion, resident philosopher and researcher. Priestley agreed; the arrangement lasted seven years.

It was at Shelburne's Bowood House that Priestley made his greatest discovery—one that led to modern chemistry.

Before I get to that, however, I'll stop writing and get something to eat.

I must here tell you, Dr. Franklin, that I occasionally despair that you've not yet been found. The exca-

Bowood house today

vations in Philadelphia continue apace. I daily check the website of the *Philadelphia Inquirer*, your hometown's most prominent newspaper, and read that construction workers have unearthed many ancient objects—dinner plates, clay pipes, silverware—but no oak barrel, at least none they've announced. I sometimes feel like the wife of a ship's captain whose vessel is overdue at home port. I must assume the increasing odds of your imminent discovery. But for that optimism, I couldn't continue my self-appointed mission to update you with these emails.

From: "Stuart Green" <stuartgreenmd@yahoo.com>
To: "Benjamin Franklin" <dr_benjamin_franklin@yahoo.com>
Subject: **A candle burned in this air with a remarkably vigorous flame**

Dear Doctor Franklin:

While at Lord Shelburne's estate, Joseph Priestley, in 1773, made his most important chemical discovery when he started roasting a material called red calx (now *mercuric oxide*.).

Priestley knew that red calx combined mercury with something else, so he decided to see if it contained any gases. Rather than using a charcoal fire—which itself emitted a gas—to heat the calx, Priestley came up with an ingenious way to raise the material's temperature. He first set the calx in a dish floating on liquid mercury inside an inverted bell jar. Next, he focused the sun's intense heat on the red calx using a 12-inch magnifying glass.

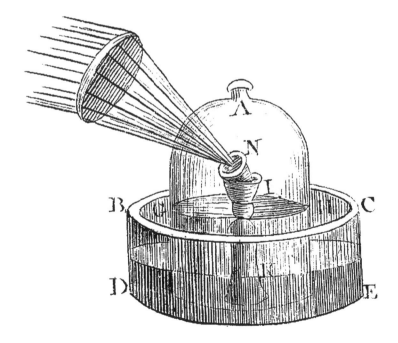

Using a burning lens (magnifying glass) to heat red calx without fire

He immediately obtained mercury and a substantial quantity of a gas. Priestley observed, however, that the gas released from red calx didn't dissolve in water. At first, he failed to understand what he had found.

Do you remember what happened next? In October 1774, Shelburne went to Paris and Priestley followed along. There he met the young chemist Antoine Laurent Lavoisier.

Priestley told Lavoisier about burning red calx with a magnifying lens, and of the curious gas he obtained. Two years earlier, Lavoisier had concluded that that combustion and calcination occur when a distinct substance in the air *combines* with a burning material rather than the material giving off something (like phlogiston). He couldn't, however, identify the gas. When he tried to reverse the process and break down a calx into its original material and a gas, the charcoal used for heating confounded his results. Thus, when he heard about the burning lens, Lavoisier suspected that Priestley had handed him the key to the puzzle of combustion. As soon as Priestley left France, Lavoisier immediately began experimenting with red calx and a magnifying lens.

Meanwhile, back in England, Priestley, after a few false turns, subjected his new gas to the candle test. As he put it, "what surprised me more than I can well express was that a candle burned in this air with a remarkably vigorous flame." Later, Priestley observed that a mouse put in his new gas, instead of dying immediately as in brewer's gas, actually survived *longer* than a mouse enclosed in a jar of ordinary room air.

Priestley, you see, had discovered what we today call *oxygen*, but he never correctly interpreted his data. I'll tell you more about this in my next email.

From: "Stuart Green" <stuartgreenmd@yahoo.com>
To: "Benjamin Franklin" <dr_benjamin_franklin@yahoo.com>
Subject: **You have set all the philosophers of Europe at work upon fixed air**

Dear Doctor Franklin:

Joseph Priestley, after observing the favorable effect his new gas had on mice and candles, soon convinced himself through additional experiments that he had discovered a gas that could absorb more phlogiston than ordinary room air, so he called the gas "dephlogisticated air." He never realized that combustion occurred when his new gas *combined* with a burning material.

Priestley kept you advised of his progress with research on the new gas, of course, but you had more pressing concerns at the time. The relationship between Great Britain and her American colonies rapidly deteriorated during the years Priestley was busily characterizing oxygen and Lavoisier probing its significance. By 1775, you had returned to Philadelphia, dismayed by Britain's inflexibility towards its North American colonies.

Six days after the opening of the Second Continental Congress on June 10, 1775, your pessimistic letter to Priestley correctly warned, "The breach between the two countries is grown wider, and in danger of becoming irreparable."

In that same communication, you'll recall, you informed Priestley that, "In coming over [across the Atlantic], I made a valuable philosophical discovery, which I shall communicate to you when I get a little time." (I assume you were referring to the Gulf Stream, the topic of a future email.)

In a letter one month later to Priestley, you seemed even more exasperated by British policy: "...we have carried another humble petition to the crown, to give Britain one more chance, one opportunity more of recovering the friendship of the colonies; which I think she has not sense enough to embrace, and so I conclude she has lost them forever."

By the time you wrote that letter, British troops had already started burning American seaports, actions you angrily characterized as "robbery, murder, famine, fire and pestilence."

Priestley obviously valued your wisdom and scientific insights. He seemed more interested in his experiments than in the war between your two countries. Anxious to apprise you of his latest discoveries, he set aside politics and wrote, "Amidst the alarms and distresses of war, it may perhaps give you some pleasure to be informed that I have been very successful in the prosecution of my experiments." He told you that he made some "observations on blood (which I believe have given great satisfaction to my medical friends) proving that the use of it in respiration is to discharge phlogiston from the system."

By this point in his research, Priestley had convinced himself that his new gas was actually the *purist* form of air, because it did everything ordinary air could do (*i.e.*, support combustion, sustain life in a bell jar) only for longer. He had, in fact, tried breathing his gas and found that it not only supported respiration, but also caused a certain lightness of this breast, as he put it, when inhaled.

Priestley subsequently proposed that the pure air he created by burning red calx with a lens might some day have therapeutic value in respiratory diseases, which is precisely why you had those little tubes in your nostrils for a time after you revived.

Blood, Priestley proclaimed, whether "congealed and out of the body" or "fluid and in the body...becomes of a florid red" in his new gas, but became dark in brewer's gas." He also advised you that the same process applied to "respiration, the calcination of metals, or any other phlogistic [burning] process." (By this time, Priestley had become aware of certain discoveries of Lavoisier that pointed to the similarity of these three events.)

You replied that, "I find you have set all the philosophers of Europe at work upon fixed air; and it is with great pleasure I observe how high you stand in their opinion; for I enjoy my friend's fame as my own."

And indeed, so have you both remained in opinion of all who followed.

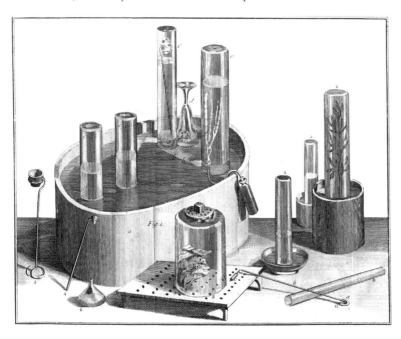

Priestley's equipment from *Observations of Different Kinds of Air*.
Note mouse in bell jar with sprig of mint.

From: "Stuart Green" <stuartgreenmd@yahoo.com>
To: "Benjamin Franklin" <dr_benjamin_franklin@yahoo.com>
Subject: **Murdered in mephitic air so many honest harmless mice**

Dear Doctor Franklin:

In early 1775, your friend Lavoisier began making claims about the gas generated by heating red calx without mentioning Priestley's prior discovery. Lavoisier reported that the gas generated from red calx was *more* combustible and more breathable than common air, although he, like Priestley, first missed the full significance of his claim.

It didn't take long for Priestley to publish a report of his own, chiding Lavoisier for the lapse. Thereafter, in all subsequent publications by Lavoisier, Joseph Priestley's name appeared prominently.

Since Lavoisier's acknowledgement of Priestley's work occurred *after* you arrived in Paris, I've wondered if you personally intervened to persuade Lavoisier to share his glory with Priestley. I imagine you had influence with Lavoisier, having spent time working with him on a couple of projects.

Within six months of your December 1776 arrival in Paris, Lavoisier, I understand, invited you to his laboratory for a demonstration. I assume he wanted to show you that Priestley's new gas from red calx combined with metals during calcination.

You were obviously impressed. Here's how you described it: "Mr. Lavoisier, the other day showed an experiment.... He kindled a hollow charcoal, and blew into it a stream of dephlogisticated air. In the focus, which is said to the hottest fire human art has yet produced, he melted platina in a few minutes."

Lavoisier had by then named Priestley's gas "principe oxygine" [acid-making principle] (He created the word oxygen because the calxes of metals burned in the gas formed acids when dissolved in water.) He later selected the name "hydrogine" [water-maker] for the "inflammable air" of Cavendish because it formed water upon combustion.

During the late 1770s and early 1780s, Lavoisier gradually snuffed out the phlogiston theory. He realized that combustion occurred when a material combined with oxygen and that, as he later wrote you, phlogiston *in Stahl's sense of the word* didn't exist.

Lavoisier, in spite of his remarkable insights into the nature of chemical reactions, mistakenly assumed that *fire* is an imponderable substance, emitted by burning materials. He thus unnecessarily substituted fire for Stahl's phlogiston in chemical reactions.

It appears you did your best to keep Priestley informed of Lavoisier's progress. Simultaneously, you described advances in chemistry to your frequent correspondent Dr. Jan Ingenhousz, the court physician in Vienna.

Ingenhousz, stimulated by Priestley's mint sprig experiments, started research on plant metabolism and physiology. He established that plants absorb oxygen and produce brewer's gas (carbon dioxide) during darkness, but in daylight they do the opposite.

I get the sense from reading your letters to Priestley that you were thoroughly disgusted with the Revolutionary War and its terrible cost in blood and capital.

Your 1782 communication with the Unitarian minister seems particularly harsh: "Men I find to be a sort of beings very badly constructed, as they are generally more easily provoked than reconciled, more disposed to do mischief to each other than to make reparation, much more easily deceived than undeceived, and having more pride and even pleasure in killing than in begetting one another, for without a blush they assemble in great armies at noon day to destroy, and when they have killed as many as they can, they exaggerate the number to augment the fancied glory."

I suspect that you assumed that Priestley, a minister, might not share your gloomy outlook about mankind because you wrote, "in your zeal for their welfare, you are taking a great deal of pains to save their souls." You then facetiously reminded your friend of the mouse experiments, suggesting that Priestley should "repent of having murdered in mephitic air so many honest harmless mice, and wish that to prevent mischief you had used boys and girls instead of them."

Priestley also hoped for speedy peace negotiations after the fighting stopped, if only to see you again: "I am exceedingly concerned to find that it is so difficult a thing to make peace...If I had any voice in the business, the prospect of seeing you once more in this country would be a strong additional motive to accelerate the negotiations."

Unfortunately, his desire, as you may recollect, never materialized.

Lavoisier dans son laboratoire

From: "Stuart Green" <stuartgreenmd@yahoo.com>
To: "Benjamin Franklin" <dr_benjamin_franklin@yahoo.com>
Subject: **Method of making salt-petre**

Dear Doctor Franklin:

In this email, I'll remind you of your productive relationship with Antoine Lavoisier that continued past his phlogiston research. You worked closely with him on two important projects, one of interest to King Louis XVI and Parisian physicians, and the other of concern to General Washington and the American Continental Congress.

Regarding another collaboration, I needn't remind you of the Royal Commission investigation of Dr. Mesmer's animal magnetism, the subject of a future email.

The second partnership involved gunpowder. When hostilities broke out between Great Britain and her North American colonies at Lexington and Concord on April 19, 1775, the so-called "shot heard 'round the world" stirred British garrisons to action. They immediately began seizing as many American gunpowder stores as they could.

Members of Congress in Philadelphia, knowing that America had no gunpowder mills, feared a war without that key ingredient. Gunpowder, as you know, consists of about 75% powdered saltpeter and 25% charcoal with a sprinkling of sulfur powder. The problem for the Americans was procuring and processing the saltpeter.

To this end, Congress commissioned a pamphlet (*Several methods of making salt-petre; recommended to the inhabitants of the United Colonies, by their representatives in Congress*) describing the manufacturing steps required to safely produce saltpeter. I know that you wrote an article in that publication called "Method of making Salt-petre in Hanover, 1766."

As soon as you arrived in Paris in December 1776, you set about procuring gunpowder and saltpeter while, at the same time, sending French chemists to America to build gunpowder works at Congress's expense. The French were, of course, pleased to sell Americans gunpowder because the explosive could ultimately macerate Great Britain, their perpetual enemy.

France, perhaps better than any nation, understood the importance of a steady supply of gunpowder. Many in that country blamed their loss to Britain of the Seven Years' War on deficient gunpowder reserves, a consequence of production limitations and profiteering.

When Louis XVI ascended to the throne in 1774, he was determined that France would never again want for gunpowder, so he appointed the efficient A.R.J. Turgot as his finance minister. Luckily for America, Turgot went to France's top chemist, Antoine Lavoisier, for advice on gunpowder production. By 1775, Lavoisier had reorganized the processing of saltpeter and gunpowder in France, assuring an adequate stock for sale to America.

Lavoisier wrote a number of pamphlets to aid producers, sellers and distributors of saltpeter and gunpowder. By spending time with Lavoisier, you learned much about making gunpowder, transmitting this information to your countrymen. In return, I assume you provided Lavoisier with concepts you developed while working with Cavendish and others on the Purfleet Magazine project.

I know as fact that you and Lavoisier co-authored two reports about safe gunpowder storage during lightning storms.

During the time you were trying to conclude a peace treaty with Britain, Lavoisier and his fellow French chemists set about to formalize chemistry in line with their anti-phlogiston concepts. Two years after you left Paris for Philadelphia in 1785, Lavoisier and his colleagues produced *Methode de Nomenclature chimique*, which established the system of naming chemicals we still use today.

Antoine Lavoisier

Chemists from other countries weren't immediately impressed with Lavoisier's system. Red calx, for instance, in the new nomenclature became mercuric oxide—a seemingly logical term. To some 18th-century wags, however, the name suggested the hide of an ox named Mercury.

Lavoisier's 1789 masterwork, *Traite elementaire de chimie* contained the complete exposition of his system. With its publication, Lavoisier realized his dream of achieving immortality in his own lifetime, although his mortal days on Earth were numbered.

Immediately upon publication, Lavoisier sent you a copy of his treatise. In an accompanying letter soliciting your backing to finally dispose of phlogiston, Lavoisier acknowledged your important place in 18th-century science. He wrote, "your support and that of some European scientists who are unprejudiced in such matters is all I care for…I would regard this revolution as much advanced or even completed if you take our side."

Dr. Franklin, did you ever commit to Lavoisier's overthrow of phlogiston? Here's one of the first questions historians of science will ask you when they get a chance: "What was your position on the *phlogiston vs. oxygenation* theories of calcination and combustion?" With my emails, you should be able to answer such a query with a modern outlook.

Lavoisier, as you know, was the scion of a family of lawyers and tax collectors, a inheritance that cost him his head during the troubled times in France that started shortly after you departed.

Joseph Priestley, in 1780, left his sheltered existence with Lord Shelburne and returned to the pulpit, this time in Birmingham. Without the support and patronage of Shelburne, Priestley's scientific research tapered off but his political writing increased. He produced two religious pamphlets (*The History of the Corruptions of Christianity; History of Early Opinions Concerning Jesus Christ*) that challenged such basic doctrines as virgin birth, the Holy Trinity and so forth. This brought him to the unfavorable attention of George III and his court.

Later, when Priestley penned a document supporting the ideals of the French Revolution and predicted future limits on the power of the British sovereign he became even more of a marked man.

When the authorities did nothing to protect his home and property during pro-monarchy riots in 1791, Priestley decided to quit England and settle in the New World. It took him a couple of years to get organized for the journey, but in 1794 (four years after your death and the same year Lavoisier met the guillotine's blade) Priestley moved to Pennsylvania with his three sons, where he founded America's first Unitarian Church

Both John Adams and Thomas Jefferson admired Joseph Priestley and paid homage to him as a scientist and man of the cloth. Jefferson consulted Priestley on the curriculum for his newly founded University of Virginia. Even after coming to America, Priestley clung to the sinking ship of Stahl's phlogiston, although in the very end he conceded that phlogiston theory functioned like a misplaced mariner's lantern, leading scientific wayfarers astray.

Benjamin Franklin meets Louis XVI, Marie
Antoinette, and the Dauphin in the Garden of the Palace of Versailles
By Andre Castaigne (1861-1929)
(Author's Collection)

From: "Stuart Green" <stuartgreenmd@yahoo.com >
To: "Benjamin Franklin" <dr_benjamin_franklin@yahoo.com>
Subject: **The power of man over matter**

Dear Doctor Franklin:

Today, I'll summarize the most important advances in chemistry during the past two hundred years. The beginning of the 19th century, as I've said, started a chemical revolution based on Lavoisier's treatise. During the French revolution, chemistry's torch of discovery passed to England.

Humphrey Davy

Humphrey Davy was born in Penzance, England in 1778 while you and Lavoisier were conferring about gas and gunpowder. When twenty-one years old, Davy became the apprentice to a physician-chemist and began a long career in science. He quickly focused attention on *electrochemistry*—the effect of passing an electric current through chemical compounds. He also pioneered *electroplating* (using electricity to deposit a thin layer of one metal on another) and contributed to a number of industrial processes. His laboratory assistant Michael Faraday started as an electrochemist and later moved to electromagnetism. As I mentioned in an earlier email, Faraday made discoveries that led to electric motors and generators.

Although many individuals illuminate our modern understanding of chemical reactions, two names glow brighter than the rest: John Dalton and Dimitri Mendeleyev. Dalton developed the *atomic theory* of matter and Mendeleyev placed those atoms in an orderly arrangement.

John Dalton

John Dalton, a Quaker scientist, shared his birth year (1766) and national origin with Edward Jenner, the conqueror of smallpox. Like other innovative chemists I've mentioned, Dalton's early work involved gases. He established a system of writing chemical symbols to replace the alchemists' zodiac-based notations. His 1808 masterwork *New System of Chemical Philosophy* gave us our modern concept that each element is composed of extremely small identical particles, similar in certain characteristics to other elements, but differing enough to give each element its unique properties. Dalton's *atoms* were indestructible: they combined in fixed proportions to form *compounds*.

Dalton, known for his simple Quaker clothing and retiring bachelorhood, must have inspired as a teacher; his visionary ideas influenced generations of scientists. John Dalton proved a worthy successor to Antoine Lavoisier in modern chemistry's saga.

During the thirty years after Dalton proposed his atomic theory, Dr. Franklin, 19th-century chemists identified new elements at the astounding rate of one per year. More significantly, chemists perceived an organized relationship between the elements, with some having properties similar to others. Several thinkers tried to make sense of it all, but none so successfully as Dimitri Mendeleyev, a Siberian-born chemist.

Dimitri Mendeleyev

Mendeleyev, like yourself the last boy of a large family, received his education in St. Petersburg and became a professor of chemistry there. Lacking a modern Russian chemistry textbook, Mendeleyev accumulated everything known about the subject soon after he began teaching.

Like others before him, Mendeleyev recognized kindred features among the 63 known elements. He grouped them into columns and rows based on how they combined to form compounds (especially salts), the similarities of their physical properties, their specific volumes and atomic weights. His

Periodic Table of Elements (1869), unlike competing constructs, contained squares where no known element existed at the time, but were eventually found.

Today, Mendeleyev's chart hangs in every chemistry classroom on the planet, the ultimate tribute to the Siberian's remarkable insight into the organization of matter.

As the 21st century commences, gas chemists will continue to play a vital role in our wellbeing. Recall that animal respiration experiments provided essential information about whether a newly discovered gas was mephitic (poisonous) or wholesome, to use your terminology.

In the future, *exhaled-breath analysis* will eliminate most blood tests now used by doctors. Although the amounts of volatile substances in expired breath—reflecting their levels in blood—are extremely minute, they can be detected with specialized but costly equipment. Once the price of such devices comes down, the number of needle-sticks will correspondingly diminish and diagnostic bloodletting of our time will go the way of therapeutic bloodletting of yours.

In 1780, you wrote Priestley a letter that displayed your oft-professed faith in the value of scientific pursuits:

> The rapid progress true science now makes, occasions my regretting sometimes that I was born so soon. It is impossible to imagine the height to which may be carried in a 1000 years the power of man over matter. We may perhaps learn to deprive large masses of their gravity & give them absolute levity, for the sake of easy transport. Agriculture may diminish its labour & double its produce. All diseases may by sure means be prevented or cured, not excepting even that of old age, and our lives lengthened at pleasure even beyond the antediluvian standard.

If the development of morals made comparable advances to those of science, you concluded, "Men would cease to be wolves to one another, and that human beings would at length learn what they now improperly call humanity."

It hasn't happened yet, Dr. Franklin.

From: "Stuart Green" <stuartgreenmd@yahoo.com >
To: "Benjamin Franklin" <dr_benjamin_franklin@yahoo.com>
Subject: **On the bad effects of lead**

Dear Doctor Franklin:

Your July 31, 1786 letter to London friend Benjamin Vaughan has, for the past thirty years, added fuel to a burning public debate about the toxic effects of lead. Considering your interest in lead's poisonous consequences, I thought you might want to hear about our present knowledge on the subject.

Although direct exposure among lead workers has lately decreased because of favorable workplace changes, environmental lead levels gradually rose during the two hundred years after your letter to Vaughan.

In case you've forgotten, you penned the Vaughan letter about lead's dangers after you arrived in Philadelphia from your eight-year diplomatic mission to France. Your final encounter with the Jamaica-born plantation owner occurred during a particularly difficult time in your life.

Everyone knows of your troubled relationship with your son William. After all, he remained a loyalist during the Revolutionary War whereas you inflamed the opposition. I realize that you two met on your homeward-bound ship at Southampton after the war, but never reconciled.

Those two days at anchor must have seemed a mixed blessing. On the one hand, you had the pleasure of seeing old British friends like Vaughan for a final farewell, but the strained encounter with William no doubt dampened your happiness. (I apologize, Sir, for reawakening painful memories; I simply hoped to refresh your recollection of the Vaughan encounter.)

Your Vaughan letter contained a chronology of your experiences with lead, starting when you were a just young lad. Recall telling Vaughan that, as a boy in Boston, you learned that people in North Carolina said that New England rum "poisoned their people, giving them the dry bellyach, with a loss of the use of their limbs." Boston authorities visited distilleries and found several using lead in fabricating rum barrels. The Massachusetts legislature soon prohibited the practice because "the physicians were of the opinion that the mischief was occasioned by that use of lead."

Next, you wrote of your experience when, aged 18, you went to London to buy a printing press, type, paper, and other supplies needed to set up a print shop in Philadelphia. Since a promised letter of credit from Pennsylvania's Royal Governor William Keith never materialized, you toiled as a print shop compositor in London to earn money for passage home.

While at the shop, you noticed printers "drying a case of types…by placing it sloping before the fire." Obviously, the practice appealed to you, because the warm lead pieces kept your hands comfortable in cold weather.

But, as you told Vaughan, "an old workman observing it, advised me not to do so, telling me I might lose the use of my hands by it." You went on to say, "This, with a kind of obscure pain that I had sometimes felt as it were in the bones of my hand when working over the types made very hot, induced me to omit the practice."

You were understandably curious, it appears, and wanted to know more, so you shrewdly asked the owner of a type foundry about the disorder. As often happens when querying a supervisor about occupational injuries, the workers were blamed for the problem: "...he made light of any danger from the effluvia [which we now call *vapors*], but ascribed it to particles of the metal swallowed with their food by slovenly workmen, who went to their meals after handling the metal, without well-washing their fingers, so that some of the metalline particles were taken off by their bread and eaten with it."

You, however, weren't entirely convinced: "the pain I had experienced made me still afraid of those effluvia."

The more you looked into the matter, the more you worried. For instance, in Derbyshire you heard about lead smelting operations there and learned that "the smoke of those furnaces was pernicious to the neighboring grass and other vegetables. But I do not recollect to have heard any thing of the effect of such vegetables eaten by animals."

You were evidently impressed by the numbers of patients at La Charité Hospital suffering from symptoms of lead exposure because you told Vaughan that, "all the patients were of trades that some way or other use or work in lead; such as plumbers, glasiers, painters, &c."

We now call lead toxicity *plumbism*, from the Latin for lead (plumbum). Lead-lined pipes traversed ancient Rome, carrying drinking water to the ruling class. Certain historians claim that the Roman Empire's collapse followed low-level lead poisoning of the capital's aristocrats.

In the 1600s, unscrupulous wine merchants sweetened their products with lead salts, which caused abdominal cramping—initially thought related to wine's acidity. By your times, however, the relationship between lead and bellyaches (dry-gripes as you called it) must have been apparent.

When you learned that your friend Joseph Galloway suffered cramping, you astutely told him, "The dry gripes are thought here to proceed always from lead taken by some means or other into the body. You will consider whether this can have been your case, and avoid the occasion. Lead used about the vessels or other instruments used in making cyder, has, they say, given that cruel disease to many."

What you told Vaughan about inhibition of moss growing on the sides of houses struck me as a particularly interesting observation: "if there be any thing on the roof painted with white lead... there is constantly a streak on the shingles from such paint down to the eaves, on which no moss will grow, but the wood remains constantly clean & free from it." You went on to mention a European family that has such a roof; they all had "dry-bellyach, or colica pictonum, by drinking rain water."

Gutenberg made his movable type out of lead, allowing easy recasting—the secret of his success as you, a printer, know better than most.

By the middle of the 17th century, occupational exposure to lead among foundry workers was recognized as a cause of paralysis. Bernardino Ramazzini, an early pioneer in lead studies, characterized occupational lead poisoning thus, "First their hands became palsied, they then became paralytic, splenetic, lethargic and toothless..."

Lead casting house, from Humphreys' *Nature Display'd*

Do you recall proclaiming in 1768 that you had "long been of opinion, that that distemper proceeds always from a metallic cause only, observing that it affects among tradesmen those that use lead, however different their

151

trades, as glazers, type-founders, plumbers, potters, white lead-makers and painters"?

With time, interested physicians began to understand the vague features of early lead poisoning. Plumbism's first symptoms include irritability, loss of appetite, sleeplessness, behavioral changes, headaches and abdominal pains. As lead levels in the bloodstream increase, a person experiences tingling and numbness of the limbs followed by muscle fatigue. The weakness in the upper limb is characterized by muscle atrophy and a wrist-drop effect (which you called "the dangles"). Debility in the lower extremities causes staggering. Constipation sets in when bowel muscle function decreases. Seizures and coma may ensue. Acute exposure causing a very high blood lead level may result in coma and death without the initial warning signs. (Clearly, you were wise to avoid additional exposure to lead fumes once you experienced early symptoms of lead poisoning.)

Based on the features of plumbism, it's obvious that lead damages the nervous system.

Once lead's toxic effects were recognized in the mid 19th century, one would've expected a decline in the use of lead-based products. Instead, the opposite happened. Many items today containing lead have contributed to present-day health concerns. As in your time, the problem of lead-based paints is still with us today, especially in older homes where young children sometimes eat flaking-off paint and inhale paint dust. (Lead interferes with brain development during growth.)

A modern device used to store an electric charge is the *lead-acid battery*; burning them increases atmospheric lead.

With the exception of lead-acid batteries (which are safe when properly handled), the 20th century has witnessed a sharp decline in the use of lead-based products as public health officials became increasingly convinced of lead's dangers, even at very low levels. Each step along the legislative pathway, however, lead industry trade groups did their best to block progress. They supported research of dubious value that minimized the hazards of environmental lead.

Lawsuits by individuals claiming harm by lead products have proven an effective, but often arbitrary, way to force needed changes.

You were prophetic, Sir, when you pointed out to Benjamin Vaughan that, "the opinion of this mischievous effect from lead, is at least above sixty years old; and you will observe with concern how long a useful truth may be known, and exist, before it is generally received and practiced on."

Indeed, Dr. Franklin, that message has finally been received and practiced on, although it's taken more than two hundred years to do so.

NUTRITION & DIETARY SCIENCES

I had formerly been a great lover of fish   155

Animal heat arises by or from a kind
of fermentation   159

Keep out of the sight of feasts and banquets   161

Eat not to dullness   163

dr f   167

Do not make an expensive feasting   169

White as your lovely bosom   171

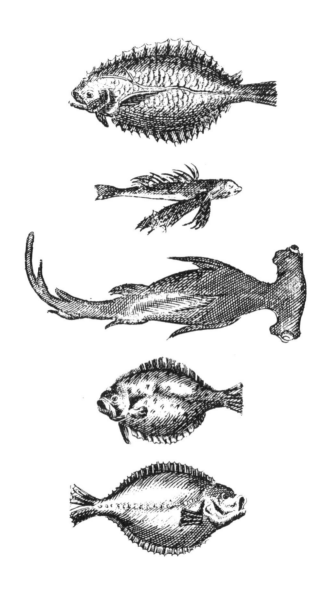

Fishes from Diderot's *Encyclopedie*

From: "Stuart Green" <stuartgreenmd@yahoo.com>
To: "Benjamin Franklin" <dr_benjamin_franklin@yahoo.com>
Subject: **I had formerly been a great lover of fish**

Dear Doctor Franklin:

"A full belly makes a dull brain," *Poor Richard* warned in 1758—sound counsel, even today. You never hesitated to offer dietary guidance to relatives, friends and strangers alike, whether asked for or not. To a modern reader, Sir, your pronouncements contain solid—and still valid—suggestions about avoiding overeating and judicious living, mixed with old wives' tales about which foods cure certain illnesses.

In a letter to Debbie, for instance, you recommended this anti-depression formulation: "Eat light foods, such as fowls, mutton, &c. and but little beef or bacon, avoid strong tea, and use what exercise you can; by these means, you will preserve your health better, and be less subject to lowness of spirits."

Likewise, to prevent heartburn, *Poor Richard* advised subscribers to "eat no fat, especially what is burnt or oily; and neither eat or drink any thing sour or acid. To cure it, dissolve a thimble-full of salt of wormwood in a glass of water, and drink it." Elsewhere in the *Almanack*, you proclaimed: "Be temperate in wine, in eating, girls, and sloth; or the gout will seize you and plague you both."

Six years later, the gout seized *you*, so what can we infer about *your* behavior?

I know that food and diet had been a long time concern of yours. Like many young people today, you dallied during your youth with vegetarianism. I enjoy reading about your entry into—and your exit from—the non-meat regime. As your *Autobiography* described the episode: "When about 16 years of age, I happened to meet with a book, written by one Tryon recommending a vegetable diet. I determined to go into it."

(I assume you're referring to Thomas Tryon's 1619 book that forewarned: "Refrain at all time such foods as cannot be procured without violence and oppression." He hinted at eternal punishment for those who killed animals for food: "Desire not variety of meats nor drinks for fear the soul be overwhelmed in the dark clouds of wrath and sorrow.")

You obviously followed Tryon's proposal, Dr. Franklin, and were chided for it by James and your fellow apprentices at the boarding house table. The harassment, I suspect, fostered dining alone. Thus you "proposed to my brother, that if he would give me weekly half the money he paid for my board I would board my self." When your brother agreed, you found that by eating simple food prepared according to Tryon's recipes, you could save half your meal expenses, leaving you "additional fund for buying books."

Your "light repast" made by "boiling potatoes or rice, making hasty pudding" or perhaps consisting of "a bisket or a slice of bread, a handful of raisins or a tart from the pastry cook's, and a glass of water" sounds unpalatable to me, quite frankly. Nevertheless, small meals, you claimed, allowed more time for study where you "made the greater progress from that greater clearness of head and quicker apprehension which usually attend temperance in eating and drinking."

You claim you continued the vegetarian diet for a year or so until you escaped from James' service and headed south by sea towards New York to look for work. The boat, however, "becalmed off Block Island." Your fellow passengers "set about catching cod and hawled up a great many." Up to that point, you wrote, "I had stuck to my Resolution of not eating animal food; and on this occasion, I considered with my master Tryon taking every fish as a kind of unprovoked murder, since none of them had or ever could do us any injury that might justify the slaughter."

I've told at least a hundred people—including many vegetarians—what happened next: "I had formerly been a great lover of fish, and when this came hot out of the frying pan, it smelt admirably well. I balanced some time between principle and inclination: till I recollected, that when the fish were opened, I saw smaller fish taken out of their stomachs: Then thought I, if you eat one another, I don't see why we mayn't eat you. So I dined upon cod very heartily and continued to eat with other people, returning only now and then occasionally to a vegetable diet."

I completely agree with your conclusive statement: "So convenient a thing it is to be a reasonable creature, since it enables one to find or make a reason for every thing one has a mind to do." Thereafter, I understand, you reverted to special diets only when ill, but, in general, it seems you ate everything with considerable pleasure, although always trying to resist gluttony.

No doubt you found it difficult to avoid overeating, considering how often you ate out when away from home. Based on your invitations, historians today know both where you dined and what you consumed.

From Diderot's *Encyclopedie*

It seems you got tired of eating meat every evening—especially in light of your youthful vegetarian days—so you kept that one-week food-and-health diary while in London. Here's a page from your diary to refresh you memory:

Monday—Din'd at Club—Beef

Tuesday—at Mr. Foxcrofts—Fish
Wednesday—Dolly's Beefstake
   Felt symptoms of Cold Fullness
Thursday—Mr. Walker's—Beef
   Predicted it
Friday at Home   Mutton
Saturday Club   Veal
   Very bad at night Wine   whey
Sunday—no Dinner   Continued bad
Monday morn. Had a good night, am better
   U[rine] deposited a reddish fine Sand.

Your missives to friends and relatives about your eating habits and theirs were but one aspect, I believe, of your abiding interest in life processes. I gained insight into your thinking about these matters by reading the lengthy letters you occasionally wrote to correspondents detailing your grasp of how the human body accomplishes its most important functions—digestion, circulation, perspiration and respiration. I'll review your concepts of *digestive physiology* in my next email.

However, since we're on the subject of eating food, I'll here remind you of your amusing plan for the gas that comes out the other end. Recall that around 1780 The Brussels Academy of Science offered a prize for the most useful scientific proposal. You rose to the challenge, noting that "in digesting food, there is created or produced in the bowels of human creatures a great quantity of wind." You also mentioned that "permitting this air to escape…is usually offensive to the company, from the fetid smell that accompanies it. That all well-bred people therefore, to avoid giving such offence, forcibly restrain the efforts of nature to discharge the wind."

You pointed out that retaining the gas is "contrary to nature," and caused great pain that "occasions future diseases, such as habitual cholics, ruptures, tympanies, &c. often destructive of the constitution, & sometimes of life itself."

Reasoning that if the gas didn't have "the odiously offensive smell accompanying such escapes, polite people would probably be under no more restraint in discharging such wind in company, than they are in spitting, or in blowing their noses."

Your scientific proposal, therefore was "to discover some drug, wholesome & not disagreeable, to be mixed with our common food or sauces, that shall render the natural discharges of wind from our bodies not only inoffensive, but agreeable as perfumes." Perhaps, you suggested, "a little powder of lime (or some other thing equivalent) taken in our food, or perhaps a glass of limewater drank at dinner" might make the gas "rather pleasing to the smell."

The importance of such a discovery, you stated, would follow Sir Francis Bacon's imperative to bring the benefits of science "home to men's business and bosom." Indeed, you concluded, when compared to the value perfumed flatulence, the rest of scientific progress was "scarcely worth a **fart**-hing."

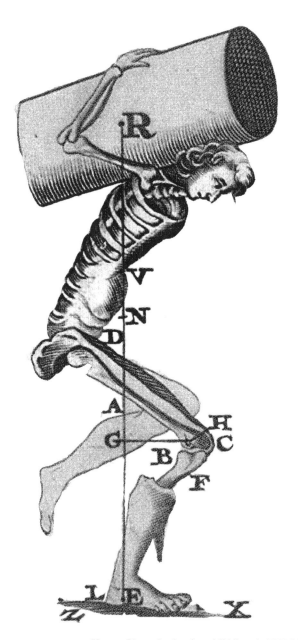

Human Biomechanics, from Middleton's 1777 *Dictionary*

From: "Stuart Green" <stuartgreenmd@yahoo.com >
To: "Benjamin Franklin" <dr_benjamin_franklin@yahoo.com>
Subject: **Animal heat arises by or from a kind of fermentation**

Dear Doctor Franklin:

Your understanding of digestion, Sir, stemmed from ideas that went back to Galen, a first-century A.D. Greek physician. His pronouncements, refined by centuries of subsequent philosophers, explained how ingested food turned into the four humors—blood, phlegm, black bile, yellow bile—that made up the bodily fluids. Imbalance among the humors caused illness.

I must admit, however, that I found numerous variations of these concepts, depending on where I looked. In most systems, the blood itself might be out of harmony, leading, for example, to phlegmatic blood or melancholic blood, both conditions adversely influencing health and well being.

To account for the fuel that sustained bodily warmth, Dr. Franklin, you correctly surmised that heat was extracted from food during a fermentation-like digestive process. In fact, some authorities recognize you as the *first* person to understand that consumption of food was a form of slow combustion. Your 1757 letter to Charleston physician John Lining garnered you such credit. In that communication you offered the following speculation: "I imagine that animal heat arises by or from a kind of fermentation in the juices of the body, in the same manner as heat arises in the liquors preparing for distillation..."

You clearly marveled that heat generated in a distiller's fermentation vat "shows by the thermometer, as I have been informed, the same degree of heat with the human body, that is about 94 or 96." You then made the following analogy: "Thus, as by a constant supply of fuel in a chimney, you keep a warm room, so by a constant supply of food in the stomach, you keep a warm body." Did this remarkable insight stem from thinking about your Pennsylvania fireplace and related speculations about heat and warmth? Please advise.

You also rhetorically asked: "Where did heat in food come from?" You had an answer for that too, surmising that the heat had been acquired by food during its growth. Here's how you put it: "I have been rather inclined to think that the fluid, fire, as well as the fluid, air, is attracted by plants in their growth...That when they come to be digested...part of the fire as well as part of the air, recovers its fluid active state again, and diffuses itself in the body..."

If I correctly understand your concept, Dr. Franklin (and I think I do), you assumed that fire—as a distinct element—became assimilated into plants during growth and was released during digestion or fermentation.

You indicated that the same thing occurred with "the fire emitted by wood and other combustibles when burning..."

Additionally, you perceived that the amount of fire incorporated in wood or coal accounted for its combustibility and that "gunpowder is almost all solid fire." In a remarkable sentence that incorporated all four Greek elements, you concluded by saying that, "what escapes and is dissipated in the burning of bodies, besides *water* and *earth*, is generally the *air* and *fire* that before made parts of the solid." [I added italics for emphasis.]

As with so much in modern science, your explanation of nutrition's energy transfer has proven correct as we acquired a deeper understanding of what actually happens during food formation and consumption: Our sun's *fire* energy is absorbed as light by plants, which convert that energy into chemical bonds, an energy storage system similar to the Leyden jar. In fact, the process functions precisely that way, using controlled "electric fire" (electron transfer) as the ultimate fuel.

We recover the sun's fire by breaking those bonds chemically, thereby releasing the stored energy.

I couldn't find in any of your writings an explanation of how you thought the stomach actually separated food into its component parts. René Réaumur, the French researcher of your era who developed a thermometer scale, looked into the matter. He determined that gastric fluid was a great solvent of food; that it was an acid; and that the stomach's food breakdown didn't involve putrification.

By the early 19th century, chemists determined that gastric fluid as *hydrochloric acid*, the same substance your friend Dr. Ingenhousz poured over iron to make hydrogen gas.

In 1833, William Beaumont, a U.S. Army surgeon, published the results of his research on a French-Canadian fur trader who, after an accidental gunshot wound, developed a fistula from his stomach to his skin, which provided gastric contents for analysis.

After Beaumont's publication, physiologists began creating fistulae in dogs for research purposes, leading to great advances in the knowledge of digestive processes. Within a few decades of Beaumont's investigations, scientists learned about chemicals called enzymes that further broke down food after it left the stomach. Our current grasp of digestion's end products, however, had to await the discoveries of 20th-century chemistry. They're still ongoing.

Chemical
Apparatus
of Jan
Ingenhousz

From: "Stuart Green" <stuartgreenmd@yahoo.com>
To: "Benjamin Franklin" <dr_benjamin_franklin@yahoo.com>
Subject: **Keep out of the sight of feasts and banquets**

Dear Doctor Franklin:

As your strength improves, you'll eventually get out onto the streets of Philadelphia where you'll notice a remarkable number of fat people waddling about. Perhaps your return will cause renewed interest in your recommendation to "eat and drink such an exact quantity as the constitution of thy body allows of, in reference to the services of the mind." Indeed, we all should heed this advice from *Poor Richard*: "Excess in all other things whatever, as well as in meat and drink, is also to be avoided."

While a younger man, you advocated daily exercise and restrained dining; you seemed particularly concerned about the effects of overeating. *Poor Richard* advised "Many dishes, many diseases" and "He that never eats too much will never be lazy."

Your warning about eating habits and temperance, especially as it relates to a full belly, certainly makes sense to anyone today who walks out of a dining room stuffed to near explosion: "That quantity that is sufficient, the stomach can perfectly concoct and digest, and it sufficeth the due nourishment of the body…The difficulty lies, in finding out an exact measure."

Your words, Sir, are truer today than in your own time because the portions served are now so large: "If thou eatest so much as makes thee unfit for study, or other business, thou exceedest the due measure." Your ideas on *due measure* appears elsewhere in your writing as well: "If thou art dull and heavy after meat, it's a sign thou hast exceeded the due measure; for meat and drink ought to refresh the body, and make it cheerful, and not to dull and oppress it."

We'd all find it easier to resist temptation if we followed your suggestion to "keep out of the sight of feasts and banquets as much as may be; for 'tis more difficult to refrain good cheer, when it's present, than from the desire of it when it is away." In fact, after the huge meal I had last night, I'm going to try your proposal to "…fast the next meal, and all may be well again, provided it be not too often done; as if he exceed at dinner, let him refrain a supper…"

The hazards of overeating seemed much on your mind, yet in spite of your Puritan upbringing in Boston and your Quaker readers in Philadelphia, you never invoked divine condemnation of gluttony, only sound recommendations any modern *nutritionist* would offer. Today's experts, for example, grumble that our mode of seasoning food stimulates overeating. Their thought is hardly new, considering this query to your club, the Junto: "Whether those meats and drinks are not the best, that contain nothing but their natural tastes, nor have any thing added by art so pleasing as to induce us to eat or drink when we are not athirst or hungry..."

Research nutritionists try to ascertain what diets are best for specific illnesses. *Nutritional maintenance*, as it's called, is deemed so important that very sick patients are given sustenance directly into their veins, bypassing the stomach and digestive system altogether. The reason for such support confirms your admonition: "when malignant fevers are rife in the country or city where thou dwelst, 'tis advisable to eat and drink more freely, by way of prevention; for those are diseases that are not caused by repletion, and seldom attack full-feeders."

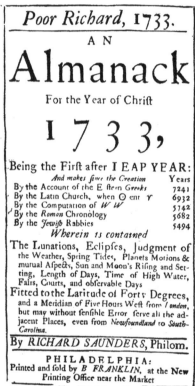

You were also centuries ahead of your time with this advice: "Use now and then a little exercise a quarter of an hour before meals, as to swing a weight, or swing your arms about with a small weight in each hand; to leap, or the like, for that stirs the muscles of the breast." Moreover, my colleagues have confirmed the value of your observation that "a temperate diet arms the body against all external accidents; so that they are not so easily hurt by heat, cold or labour; if they at any time should be prejudiced, they are more easily cured, either of wounds, dislocations or bruises."

Finally, I should inform you that modern scientists have determined that rats kept on sparse diet live longer and remain healthier than those allowed unrestricted access to food. I doubt you'll find such results surprising, considering this question posed to members of your Junto Club: "Whether it is worth a rational man's while to forego the pleasure arising from the present luxury of the age in eating and drinking and artful cookery, studying to gratify the appetite for the sake of enjoying healthy old age, a sound mind and a sound body, which are the advantages reasonably to be expected from a more simple and temperate diet."

All this talk of food must make you hungry for something that wasn't prepared in the hospital's kitchen. In my next email, I'll offer a plan to satisfy that desire.

From: "Stuart Green" <stuartgreenmd@yahoo.com>
To: "Benjamin Franklin" <dr_benjamin_franklin@yahoo.com>
Subject: **Eat not to dullness**

Dear Doctor Franklin:

If you've reached this point in our email discourse, I must assume your health has improved; your gout is under control, you're voiding freely and your respirations are with ease. As you acquire vigor you'll doubtless wish to get out and see what has transpired in Philadelphia during the past 215 years. I urge you to gain more strength —and to refresh your memory with my emails—before facing the public.

I do expect, however, that you'll soon tire of hospital food. No doubt Pennsylvania Hospital's menu has improved since that day in 1787 when, as President of Pennsylvania, a Mr. Joseph Elam petitioned you about the "scanty dinner" provided him in the Lunatic's Ward. Nevertheless, you'll not feel truly at home until you've dined on typical food of your times, and in familiar surroundings.

Although more than two centuries have passed since you last ate out in Philadelphia, it's still possible to enjoy a recognizable repast in a place well known to you—namely, the City Tavern. Recall that you spent many happy hours there with friends and acquaintances, debating issues of import, both great and small.

The original facility was damaged by fire in 1834 and razed to the ground twenty years later. In 1976, however, to help celebrate the bicentennial of the Declaration of Independence, a replica City Tavern was constructed to match, as closely as possible, the original structure. The present owner strives to bring authenticity to the bill of fare, with many offerings from your era.

The problem, Dr. Franklin, is getting you to The City Tavern without alerting the army of newspaper people who'll surround the hospital following your disinterment. After thinking about the matter, I've come up with a practical solution.

There are in Philadelphia (and other cities with a colonial past) actors who specialize in playing the role of an historic American. These individuals call themselves *re-enactors* and they don't work from scripts. Instead, they study books about their character to bring reality to the performance.

They make a living by leading tour groups around their respective towns or by appearing—for a handsome fee, I might add—at conventions or banquets. The sight of such actors strolling the streets of Philadelphia in 18th-century garb never raises eyebrows among the city's inhabitants.

I've personally dined with two of the best Benjamin Franklin re-enactors, so I can attest to their abilities.

With these facts in mind, here's my plan:

I'll ask one Franklin to convene a Philadelphia gathering of colonial American re-enactors when you've the strength to move around more sprightly. I'd expect about a dozen to show up. As part of their visit to the City of Brotherly Love, I'll arrange a tour of Pennsylvania Hospital, conducted by the facility's helpful archivist Stacey Peeples. (Typically, the visit lasts about one hour.) The tour always ends in the Hospital's Historic Library, a repository for rare 18th-century medical books and anatomic specimens.

Near the Library is a small privy. The second Benjamin Franklin bears a striking resemblance to you in both face and figure—or perhaps I should say in both countenance and corpulence.

Rest assured, Dr. Franklin, an exchange of clothing between you two in the privy need take no more than a minute. The total number of conspirators to the plot: three. The rest of the entourage will be intently interested, I'm sure, in the wax medical specimens and won't notice the switch.

After a swift substitution, you'll be on you own, Sir, impersonating a Benjamin Franklin impersonator.

The City Tavern

The group will next ambulate over to The City Tavern for a prearranged lunch. You'll feel especially comfortable surrounded by your companions Thomas Jefferson, George Washington, John and Abigail Adams, Robert

Morris, Samuel Adams, Benjamin Rush and many others as well. But mind what you say, Sir: I recommend silence, or nearly so except, perhaps, for pleasantries.

Be particularly careful not to inquire of yellow fever epidemics or brutal savages or unpaid Revolutionary War debts. Do not, under any circumstances, query Washington or Jefferson about their slaves. Instead, listen quietly to their conversations; you'll hear one word repeatedly—*gig*—meaning a paying engagement. They'll complain continuously about their agents and their lawyers and their competition.

As you walk up 8th Street and onto Walnut, look around in a seemingly disinterested manner. You'll notice tall buildings towering over the State House (now called *Independence Hall*). They all have steel frames, made possible by Sir Henry Bessemer's 1855 process that recreated for mass production ancient Chinese steel-making techniques. You must decide for yourself if Philadelphia has improved in appearance since your last hours there.

Henry Bessemer

The horseless carriages that rapidly pass through the streets are descendants of John Ramsey's steam-powered boat. As president of the Ramseyian Society, you'll appreciate a future email detailing the evolution of such contrivances.

I recommend that you exercise extreme caution upon entering the Tavern. Avoid declaring "that stairway is too close" or "the door hinges are wrong" lest you reveal yourself.

At the dinner table, you should follow your own advice; "Eat not to dullness. Drink not to elevation." Also, remember what you said about silence: "Speak not but what may benefit others and yourself. Avoid trifling conversation."

After you've supped, Dr. Franklin, the re-enactors will travel to the corner of 5th and Arch, there to partake in a Philadelphia ritual—tossing a penny on the grave of Benjamin Franklin.

This'll be your most difficult moment. I doubt you'd willingly share in such a fraud, but you can't avoid it. I suggest you shrug your shoulders to indicate you've no money. If a compatriot offers a penny, take it and follow *Poor Richard's* advice; "At a great pennyworth, pause a while." Study the disc, but not for too long. The face profile represents Abraham Lincoln, our sixteenth President, and the obverse is his memorial structure. (Lincoln emancipated Southern slaves by presidential decree.)

At that point, Sir, the choice is yours: either participate in the deception and fling the coin, or exclaim for all to hear, "*Poor Richard* says 'A penny saved is a penny got' " and pocket the piece of copper.

When finished, your group will return to the hospital to hear my lecture, "Benjamin Franklin and the Mesmerism Investigation." When the image of Marie Antoinette appears, her décolletage will distract the audience. Move quickly to room's privy and reverse the exchange. Return to your room, and it will be done.

Don't concern yourself about missing the latter half of my speech; you already know how it ends.

dr f
have 2 go 2 wedding
i email u nxt wk
s (~_~)

Sarah (Sally) Franklin Bache

William Franklin

From: "Stuart Green" <stuartgreenmd@yahoo.com >
To: "Benjamin Franklin" <dr_benjamin_franklin@yahoo.com>
Subject: **Do not make an expensive feasting**

Dear Doctor Franklin:

   I'm today in *Florida*, the former Spanish Colony, now a part of the United States. The wedding ceremony I mentioned last week will occur tomorrow. I expect a lavish party since the bride's father is vice president of the company that owns the hotel. I doubt he'll heed the kind of advice you sent wife Deborah in June 1767 when of informed your daughter's forthcoming marriage to Mr. Richard Bache: "I would only advise that you do not make an expensive feasting wedding but conduct every thing with frugality and economy...For since my partnership with Mr. Hall is expired, a great source of our income is cut...."

Deborah Franklin

I must apologize for the brevity of my last email—actually, a *text message*. I sent it from my *cell-phone*, a wireless descendent of the contrivance invented by Mr. Bell (mentioned in my email "electricity and magnetism"). We each carry one around in a pocket, powered by a miniature version of M. Volta's electric storage battery. We can either speak to distant acquaintances or, if we must remain quiet—in a classroom, for example—send a text message. Commercial enterprises have control over the means of communication (just as you ruled mail routes in your time). The companies charge their customers <u>per character</u> to send text messages, so we've developed a phonetic shorthand, similar in some ways to the phonetic alphabet you created to teach English easily to the illiterate.

We also use punctuation marks to make little faces (*emoticons*, from "emotion-icons") within the text to convey our joy or sadness or curiosity with :-) or :-(or (*_*).

In case you haven't figured it out, my text message reads:

[I] have to go to [a] wedding

I [will] email you next week.

S[tuart] [sad emoticon].

Volta's batteries

Tomorrow's the wedding. I'll email you next after I return home.

From: "Stuart Green" <stuartgreenmd@yahoo.com>
To: "Benjamin Franklin" <dr_benjamin_franklin@yahoo.com>
Subject: **White as your lovely bosom**

Dear Doctor Franklin:

I just got back from my nephew's wedding. While there I observed something so astounding that I wanted to share it with you as soon as I returned to my computer.

The marriage ceremony itself was lovely; the party after the nuptials, lavish beyond imagination. The orchestra, however, played music that, by its amplified volume, could seemingly burst eardrums. (Musicians now can intensify sound electrically, using a microphone—described in an earlier email—and cone-shaped paper *speakers* the size of dinner plates. An electromagnet vibrates the cone's surface, creating sound waves of great power.)

Not all advances in technology, Dr. Franklin, have proven suitable to everyone's taste. After an hour of ear-splitting music, I sought relief by escaping from the banquet hall. I stepped out into the tropical evening and strolled along the avenue, noting the attire worn by Florida's citizens.

As you'll soon discover, clothing styles have changed significantly from your time to mine, characterized by two trends: simplicity and immodesty. Women's dresses illustrate the point. During your era, a lady's hemline nearly reached the floor. Over the past two centuries, the bottom edge of dresses gradually ascended until now, for some women, a skirt barely covers their buttocks!

Bathing costumes have also shrunk in size; many hardly exist at all. In France, women don't cover their bosoms on beaches. (I expect you'll soon want to revisit that country!)

Speaking of bosoms, I promised you something astounding, so here it is: While walking away from the wedding reception, I spotted a theater across the street. The show that evening featured *strip-teasers*—young ladies, professional dancers actually, who remove their clothing piece by piece in time to music. (Local laws govern how far they go.)

To entice passers-by into the establishment, two scantily clad performers sat in front the theater flaunting their assets, if you catch my meaning. At that moment, Sir, I observed the cumulative miracle of modern medicine, the zenith of two centuries of enlightened thought, the end product of advances in chemistry, biology, surgical technology, and fashion evolution.

Each woman wore a low-cut blouse to better display her surgically enlarged bosoms. That's right Dr. Franklin, *surgically enlarged bosoms*. During *augmentation mammoplasty*, as it's called, surgeons permanently implant a leakproof bag containing a jelly-like substance under each mammary gland, pushing it forward. The bags come in several sizes to provide the desired

171

shapeliness. Although still far away, I could see that each of these women had selected the largest augmentation bags possible. I said to myself, "Old Ben must see this himself someday." (Pardon the familiarity, Dr. Franklin; I hope you'll understand my agitation.)

I remembered your interest in such matters. You observed how flat-chested were French women, attributing the situation to the practice of upper-class mothers sending their babies to wet nurses. Here's how you put it: "A surgeon I met with here excused the women of Paris by saying seriously that they could not give suck…and had me look at them, and observe how flat they were in the breast; they have nothing more there, says he…I have since thought that there might be some truth in his observation, and that possibly Nature finding they made no use of bubbies has left off giving them any."

Erasmus Darwin

By assuming that disuse of a body part causes inherited atrophy, you expressed the notion that biological evolution followed a "use it or lose it" principle. Your friend Erasmus Darwin's famous 4-volume poem *Zoonomia* supported this view, as did the writings of Jean-Baptiste Lamarck, protégé of your acquaintance the Comte de Buffon. In fact, today we call such a proposal *Lamarckism*—proven incorrect by research a century later. (I'll have more about this topic in a future email since you played a role in the story.)

I recall your writing to your sister's friend Catherine Ray, comparing her to winter weather: "Your favours come mixed with the snowy fleeces which are pure as your virgin innocence, white as your lovely bosom—and as cold."

Upon arrival at the inflated foursome, I observed that their owners appeared considerably older than I'd assumed from a distance. Their painted faces, caked thick with rouge, reminded me of your comment about the "fair women at Paris": "As to rouge, they don't pretend to imitate nature in laying it on."

Sir, the strip-tease dancers of Florida don't pretend to imitate nature either.

EXERCISE SCIENCE & PHYSIOLOGY

✉ Running, leaping, wrestling and swimming   177

✉ *The Cause of Heat of the Blood*   181

✉ The globules of the blood   185

✉ The frame of the auricles or ventricles
of the heart   191

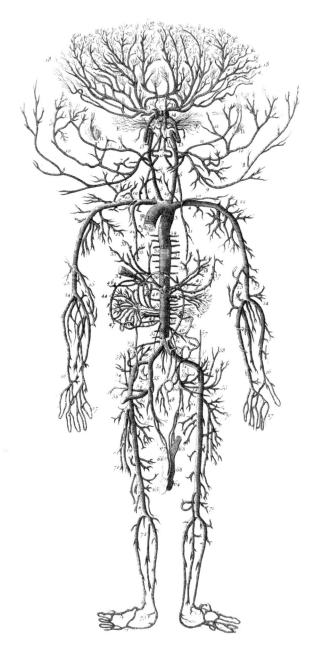

The circulatory system, from Middleton's *Dictionary*

From: "Stuart Green" <stuartgreenmd@yahoo.com>
To: "Benjamin Franklin" <dr_benjamin_franklin@yahoo.com>
Subject: **Running, leaping, wrestling and swimming**

Dear Doctor Franklin:

I assume that hibernating in a barrel for more than two centuries has made you a bit stiff. I'm confident that once your gout and bladder stone no longer limit your movements, walking will reinvigorate you.

I know you enjoyed exhilarating activity during your early years. In fact, you'll be pleased to learn that posterity has bestowed upon you an unusual honor—that of an acknowledged sportsman. In 1968, you were inducted posthumously into the International Swimming Hall of Fame.

Everyone knows that while in London at age 18, you giving swimming lessons to a young man named Wygate. You "taught him, and a friend of his, to swim at twice going into the river, and they soon became good swimmers."

Recall also demonstrating your skills to a group of gentlemen: "I stript and leapt into the river, and swam from near Chelsea to Blackfryars [a distance of 3.5 miles], performing on the way many feats of activity both upon and under water, that surprised and pleased those to whom they were novelties." You impressed the men with your swimming ability and were "much flattered by their admiration."

Later, the prominent Londoner Sir William Wyndham told you he had "two sons about to set out on their travels; he wished to have them first taught swimming; and proposed to gratify me handsomely if I would teach them...from this incident I thought it likely, that if I were to remain in England and open a swimming school, I might get a good deal of money. And it struck me so strongly, that had the overture been sooner made me, probably I should not so soon have returned to America."

(I sometimes think about how different America's history might have been if, instead of returning home, you remained in London as proprietor of Benjamin's Swimming School!)

Although you had a particularly enlightened view about the benefits of exercise, your era was filled with misconceptions about how muscles work, based on errors traceable to the ancient Greeks. Aristotle, for one, believed that airflow caused muscle movement—just as we make a balloon larger or smaller by blowing or sucking air.

I'm surprised, Dr. Franklin, that this notion persisted for 1500 years because muscles don't hiss out gas when punctured. Later, blood replaced air as the presumed force of muscle contractions.

The Italian anatomist Andreas Vesalius, in 1534, correctly discerned that a muscle moves a body part by contracting, but resumes its resting position by the action of opposing muscles. Moreover, Vesalius recognized that nerves served as messengers of "animal spirits" that made the muscles work.

Galvani's 1780 twitching muscle discovery led to a search for the way muscles are controlled

We now know that individual muscle cells contain two parallel substances, one rope-like and the other similar to centipedes. When stimulated by electricity, the centipede-like structures literally crawl along the ropes, contracting the muscle. When the centipede-like structures release their grip on the rope, the muscle relaxes.

You clearly saw value in exercise. To remind you, Dr. Franklin, here's what you said about shipboard voyages: "Want of exercise occasions want of appetite, so that eating and drinking affords but little pleasure." Many people today overcome this problem by exercising on ships during long sailings.

The prescription you offered a doctor friend sounds like something a modern physician might say: "Do you take sufficient bodily exercise? Walking is an excellent thing for those whose employment is chiefly sedentary."

In a letter to Thomas Percival about the origin of colds, you sounded like a present-day exercise promoter: "Some unknown quality in the air may perhaps sometimes produce colds, as in the influenza: but generally I apprehend they are the effects of too full living in proportion to our exercise."

Remember when your son William (in 1772 when you and he were still on speaking terms) wrote that he planned to exercise more often. You supported his decision: "The resolution you have taken to use more exercise is extremely proper, and I hope you will steadily perform it. It is of the greatest importance to prevent diseases; since the cure of them by physic is so very precarious."

You're right: the quantity of exercise should be judged "not by time or by distance, but by the degree of warmth it produces in the body."

Not long ago, America's leading muscle researchers labeled you a pioneering *exercise physiologist* because, in your letter to William, you estimated the comparative value of different kinds of exercise: "there is more exercise in one mile's riding on horseback, than in five in a coach; and more in one mile's walking on foot, than in five on horseback; to which I may add, that there is more in walking one mile up and down stairs, than in five on a level floor." They were especially pleased when you quantified your statement by using a dumb bell: "by the use of it I have in forty swings quickened my pulse from 60 to 100 beats in a minute, counted by a second watch."

Portraits depict you as a portly fellow, Dr. Franklin, but most were painted after gout limited your mobility. As a younger man, you obviously got plenty of vigorous exercise. In your *Autobiography* you bragged that, while working as a twenty-year old printer in London, you "carried up and down stairs a large form of types in each hand, when others carried but one in both hands."

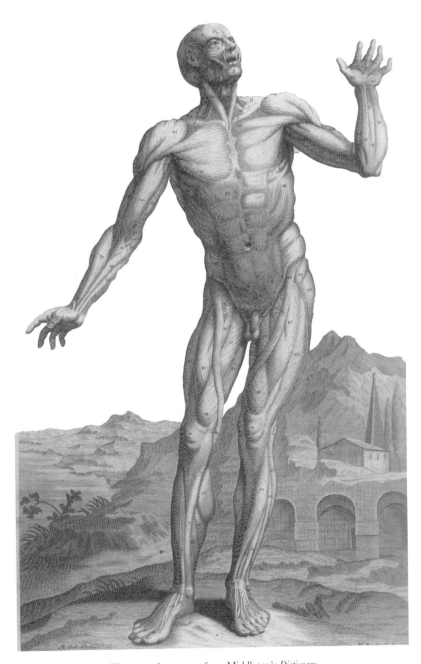

The muscular system, from Middleton's *Dictionary*

Clearly, you realized the value of exercise by the time of your 1749 *Proposals Relating to the Education of Youth in Pensilvania*, which offered numerous reasons why Philadelphians should support a secular academic institution.

The school would differ from existing North American colleges, all dedicated to training clergymen. Instead, the Philadelphia Academy (now the *University of Pennsylvania*) would educate young men with skills for business and commerce.

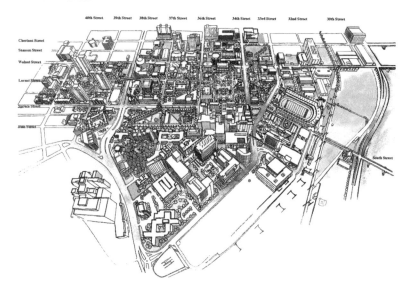

University of Pennsylvania campus today,
downloaded from the university's website.

I wonder what today's enrollees at "Penn" think of your proposal that students eat "plainly, temperately, and frugally." Considering the importance of sports activities at Penn, most would endorse your recommendation that "to keep them in health, and to strengthen and render active their Bodies, they be frequently exercised in running, leaping, wrestling, and swimming, &t."

Moreover, you wanted students "to be free from the slavish terrors many of those feel who cannot swim, when they are obliged to be on the water even in crossing a ferry."

Today, Sir, you'd get three vigorous huzzahs from college athletic directors for your *Proposals* because you maintained that "corporal exercise invigorates the soul as well as the body…For this, and other reasons, certain hardy exercises were reckoned by the ancients an essential part in the formation of a liberal character; and ought to have their place in schools where youth are taught the languages and sciences."

From: "Stuart Green" <stuartgreenmd@yahoo.com>
To: "Benjamin Franklin" <dr_benjamin_franklin@yahoo.com>
Subject: **The Cause of Heat of the Blood**

Dear Doctor Franklin:

If you're reading this email, you've the distinct honor of being the first person resuscitated after drowning in alcohol. You're also unique in another category: you're the first individual to experience *poikilothermia*—changing body temperature to match surrounding conditions. Lizards, snakes, insects, and many other "cold-blooded" creatures experience variations of internal warmth throughout the day and night. Humans and other mammals maintain a constant body temperature—*homeothermia*—regardless of the surrounding conditions.

By spending two centuries in a buried oak barrel you went through ranges in body temperature that no human had previously endured. One might call it poetic justice that your body's warmth varied so much, considering the interest you displayed in homeothermy before your encaskment.

As we discussed in an earlier email, you correctly concluded that food consumption shares features with both fermentation and combustion. You also assumed that the *heart* produces the body's warmth. You thus differed with most of you contemporaries, physicians included, who believed that friction of blood against itself generates our naturally high temperature.

I was astonished by the sophisticated level of analysis you expressed in your 1754 letter to fellow experimenter Cadwallader Colden about various aspects of bodily function. Recall that you appended to that letter your "*A Guess at the Cause of Heat of the Blood in Health and of the hot and cold fits of some Fevers.*"

You started out by debunking the commonly held idea that blood's friction with itself during circulation caused the body's heat. You insightfully wrote, "the parts of the fluids are so smooth, and roll among one another with so little friction, that they will not by any (mechanical) agitation grow warmer." You pointed out that shaking a half-full bottle of water causes neither the fluid nor the bottle to heat up. From this you drew the obvious inference: "Therefore, the blood does not acquire its heat either from the motion and friction of its own parts, or its friction against the sides of its vessels."

After reminding Colden that the heat of friction comes from the motion of "parts of solids" against each other, you declared, "The heart is a thick muscle, continually contracting and dilating near 80 times a minute; by this motion there must be constant intrafriction of its constituent solid parts; that friction must produce a heat, and that heat must consequently be continually communicated to the perfluent blood." Moreover, if I understand your view, every contraction of the heart distends the arteries, which subsequently contract again, also resulting in heat production.

The heated blood, in your conjecture, transferred its warmth to the extremities, "by this means its heat is gradually diminished, it is returned again to the heart by the veins for a fresh calefaction [heating]."

Cadwallader Colden

Your speculation here wasn't quite correct. While it's true that the heart, being a muscle, generates heat while pumping, the elastic vessel walls do not. In fact, the contribution that the heart makes to heat production is small when compared to the mass of muscles elsewhere in the body that contributes to warmth. Shivering while cold, for instance, is an involuntary attempt to raise body temperature. Such rapidly alternating movement of muscles does increase body heat and maintain survival, but not by mechanical friction, as you thought.

Forgive me for saying so, but you compounded your error in physiology when you considered the basis of perspiration. You did, however, correctly surmise that "every particle of the *Materia Perspirabilis* [now *sweat*] carries off with it a portion of heat"; but you wrongly assumed that blood became viscous during illness and that "the same viscidity in the blood and juices checks or stops the perspiration by clogging the perspiratory ducts."

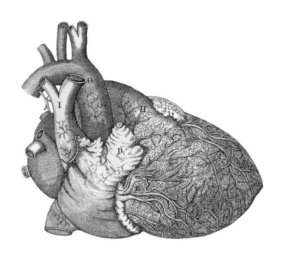

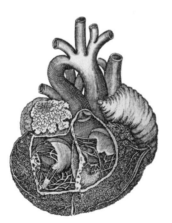

Heart from Middleton's *Dictionary*

If I may restate your concept of bodily heat dissipation, Dr. Franklin, increased viscosity of the blood prevents sweating, which, in turn, causes the body's temperature to rise, resulting in fever. That fever raises the *core body temperature*. Do I understand you correctly? As you put it, "the inward parts grow very hot, and by contact with the extremities, communicate that heat to them." The fever, in your hypothesis, leads to recovery because, "The glue of the blood is by this heat dissolved, and the blood afterwards flows freely as before the disorder."

Although your notions about the source of constant body temperature were incomplete, they did reasonably account for certain important observations. First, individuals suffering from *heatstroke* demonstrate both a rapid rise in body temperature and a cessation of sweating, as you know. Your 1757 letter to John Lining provided a perfect clinical description of heatstroke.

Recall writing that reapers in Pennsylvania, when working in the sun, can continue to "go through that labour without being much incommoded by the heat while they continue to sweat…but if the sweat stops, they drop, and sometimes die suddenly, if a sweating is not again brought on by drinking…"

Second, in that same letter you offered a clearly written (and correct) explanation for the purpose of sweating—cooling the body. You compared the difference between a live person and a dead one by noting that, in warm weather, when the ambient air temperature is about 100 degrees (four degrees warmer than normal body temperature which you understood to be 96 degrees but is actually 98.6 degrees), "a dead body would have acquired the temperature of the air, though a living one, by continual sweating, and by evaporation of that sweat, was kept cold."

Enough talk about hot and cold bodies for now. As usual, I'm late for work. I never explain to anyone at work why I'm so often tardy. If my coworkers learned that I was busy sending emails to our most eminent Founding Father, I'd be locked away in an insanity ward. For this reason, I create seemingly plausible excuses and hope I'm believed.

From: "Stuart Green" <stuartgreenmd@yahoo.com>
To: "Benjamin Franklin" <dr_benjamin_franklin@yahoo.com>
Subject: **The globules of the blood**

Dear Doctor Franklin:

I'll use this email to tell you how your London landlady's son-in-law played an important role in determining blood's various functions.

In your 1754 letter to Cadwallader Colden (mentioned earlier), you put on paper your thoughts about fever and its relation to blood clots. In keeping with the prevailing concepts of your time, you concluded that high fever dissolved blood clots much as heat melted metal; this improved circulation to sweat ducts, which enhanced sweating and improved health.

We now know that high fever is *beneficial* because most invasive disease-causing germs cannot reproduce at elevated body temperatures. For this reason, many modern infectious disease specialists recommend against administering temperature-lowering medications during a febrile illness caused by microbes. They often permit the high fevers to follow a natural course, provided they administer fluid sufficient to prevent dehydration, and use antibiotics to help our bodies kill the germs.

Your impression that fever dissolves blood clots—the prevailing viewpoint for more than two thousand years—was wrong. That blood could coagulate and form a firm mass seemed to the ancients an unwelcome impediment to the natural flow of an essential humor without any obvious benefit. Indeed, blood's thickening within the body was considered undesirable—a cause of illness. It wasn't until your era that anyone realized that blood coagulated to stop hemorrhage.

Although you couldn't possibly have known it at the time, the first step towards a rational explanation of blood clotting would occur, quite literally, in your own London backyard.

Recall that within four days of your July 1757 arrival in London, you found lodging at the Craven Street home of Mrs. Margaret Stevenson—a widow of your own age—and her lively and intelligent 23-year-old daughter Mary (whom you called "Polly"). It was your connection to Polly that involved you in the blood-clotting story.

You treated Polly Stevenson like daughter and she responded in kind. Over the next thirty years, your mutual communications included 154 letters, but there were undoubtedly others because both of you make references to documents not preserved. It's rumored that you hoped Polly would appeal to your son William, but that romance never blossomed.

Polly was still unmarried when you returned to London for your last mission in 1764, and remained so until her thirty-sixth year, when she married Dr. William Hewson, a surgeon and anatomist. Her marriage to Hewson, as

Craven Street (London) residence today

you know, lasted four years and resulted in three children—two boys and a girl. Sadly, William Hewson died from an infection acquired by accidentally cutting himself during a cadaver dissection. Perhaps you felt some consolation when Polly and her children, after the Revolutionary War, moved to Philadelphia to be with you and your family to the apparent end of your prior life.

William Hewson's career as a physician began when he studied medicine and surgery under the justifiably famous Hunter brothers. (Physicians everywhere know the name because of Hunter's Canal, a tunnel in the thigh muscle through which passes the limb's main artery and vein.)

John Hunter practiced with his brother William, ten years his senior, a relationship that proved contentious but fruitful. Moreover, John worked to elevate the surgical profession to a scientific pursuit based on anatomic principles learned during skillful cadaver dissections.

The Hunters established a school for surgeons; they also held public anatomical demonstrations.

Polly's future husband, after additional medical training in Edinburgh, returned to the Hunter brothers at their London school to teach anatomy and surgery, and to prepare anatomic specimens. In his spare time, Hewson studied the lymphatic system of fishes and other animals, a subject that also interested John Hunter.

Hewson's work on the nearly invisible lymph channels, in conjunction with the quality of his dissections of these structures, earned him membership in the Royal Society and, in 1770, the Copley Medal.

William Hewson and John Hunter got into a dispute about who actually made the most important observations about the lymphatic system, each claiming priority.

When Hewson married Polly Stevenson in mid-1770, he was forced out of the Hunter's school on the pretext that he could no longer be a "resident" assistant if he lived elsewhere with Polly. Hewson thereafter established his

184

own medical school on Craven Street and demanded the specimens he prepared while working for the Hunters.

John Hunter

A nasty quarrel ensued over ownership of Hewson's dissections. Both sides, I understand, asked you to mediate.

Your October 30, 1772, letter to William Hunter appears to have settled the matter. You wrote that on August 23, 1771, you created a memorandum to yourself "during the unpleasant time in which, as a common friend, I was obliged to hear your and Mr. Hewson's mutual complaints." While this memorandum implied that William Hunter owned all the dissections, your letter also reminded Hunter of a gift he promised to *you* of some spare specimens.

Although you wrote Hunter that you couldn't find his offer "in the confusion of my papers," it has since been located and now resides at the American Philosophical Society. (I'll ask Roy Goodman, a particularly helpful curator there, to email you a copy, although you could, if you wish, appropriate the original.)

Hunter's offer reads, in its entirety: "Dr. Hunter's compliments to Dr. Franklin. He has some preparations which he intends giving away, and if they would be acceptable to Dr. Franklin, he should be glad to see him any morning (except Tuesday) from 8 to 10 or 11 for a few minutes." You stated in your October 1772 letter, "I did so, and accepted them. I apprehended it to be your supposition in giving them to me, that as I had no use for them, I should probably give them to Mr. Hewson, which I immediately did."

Thus you took upon yourself the onus of a pass-through gift-giver.

William Hewson

After Hewson established his own medical school, he continued studying lymph and the lymphatic system. He soon discovered the active clotting principle in blood, which he labeled "coagulable lymph" (now *fibrinogen*).

When fresh blood sits around in a glass tube for a while, the blood cells settle to the bottom and a clear fluid remains on top. Hewson found that the blood's clotting ability resided in the upper liquid, rather than in the lower part with the cells. This finding alone assured his place in the history of medicine.

Hewson, however, went on to make several more important observations in his remaining few years. He was the first to describe *three* distinct varieties of blood cells. Employing a simple (one lens) microscope, Hewson correctly characterized the shape of red blood cell as discoid rather than spherical as was then believed. He also noted the presence of two kinds of white blood cells—roundish ones we now call *lymphocytes* and amorphous-looking *polymorphonuclear* cells.

About a hundred years would pass before researchers figured out the function of each cell type. We now know that red blood cells carry inspired oxygen to the tissues and return carbon dioxide to the lungs for discharge. The white cells play germ-killing and tissue-healing roles.

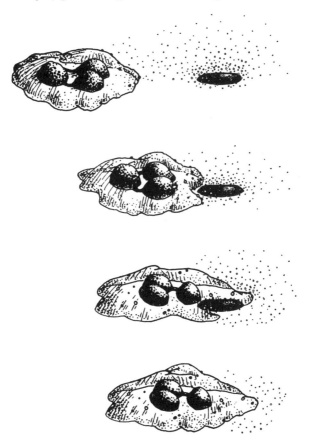

White blood cell with three-lobed nucleus (left) attacks, engulfs, and kills bacteria (right) by following gradient of chemicals emitted from bacterial surface.
From Green, S.A.: *Complications of External Skeletal Fixation: Causes, Prevention, and Treatment* (1981)

The next step in the delineation of blood coagulation after Hewson's 1771 *Experimental Enquiry into the Properties of Blood* occurred at Scotland's Glasgow Infirmary. In 1830 a young surgeon named Andrew Buchanan wrapped a large blood clot in a piece of linen and squeezed it. The liquid that oozed out made several bodily fluids clot rapidly, including the clear plasma from settled blood. Buchanan had discovered *thrombin*, which converts Hewson's fibrinogen into a fibrillar material called *fibrin*—a stringy substance that forms the mesh supporting all blood clots.

William Hunter

Subsequent discoveries about blood and clotting occurred at a steady pace for the next century and a half. Modern research into blood coagulation engages the energies of thousands of scientists and clinicians worldwide. Controlling blood's clotting activity has proven beneficial in apoplexy (strokes), heart attacks, and many other conditions as well.

In my next communication, I'll clear up certain misconceptions you had about the workings of the heart.

From: "Stuart Green" <stuartgreenmd@yahoo.com>
To: "Benjamin Franklin" <dr_benjamin_franklin@yahoo.com>
Subject: **The frame of the auricles or ventricles of the heart**

Dear Doctor Franklin:

I'd like to return to your interest in the heart's performance as a pump. By your era, heart function was understood, thanks to the 1628 discoveries of William Harvey. Recall that he conducted research on animals and used a tourniquet to analyze the arteries and veins in his own arm.

Prior to Harvey's experiments, physicians viewed the heart as a gland that secreted blood for a one-time trip to the organs. Harvey showed that the heart's right auricle takes blood returned from the body and pumps it to the right ventricle and thence to the lungs. After passing through the lungs, according to Harvey, the blood went into the left auricle and then to the large thick-walled left ventricle, which had the strength to pump blood throughout the body. Thus, blood circulated repeatedly throughout the body.

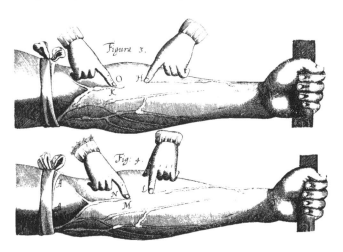

Demonstration of valves in veins
From Harvey's *De Motu Cordis*

You obviously understood and accepted Harvey's description of blood circulation. Recollect that the subject came up in a 1745 correspondence with your friend Cadwallader Colden. You put Harvey's discoveries to logical use by wondering if "the ventricles of the heart, like syringes, draw when they dilate, as well force when they contract?"

You answered your own question thus: "That this is not unlikely, may be judged from the valves nature has placed in the arteries to prevent the drawing back of the blood in those vessels when the heart dilates, while no such obstacles prevent its sucking (to use the vulgar expression) from the veins."

You misspoke, Dr. Franklin, when you mentioned valves in the arteries, because there are none. In the context of your assertion, however, could you have been writing about the *heart valves* at the bases of the great arteries, specifically the aortic and pulmonary valves, which, as you said, prevents "Drawing back" of blood into the heart when it relaxes?

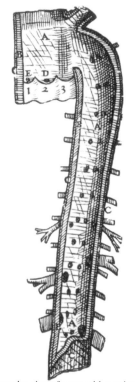

18th-century drawing of aorta with aortic valve noted

Based upon what we know about blood circulation, you incorrectly assumed that the heart "draws as well as drives the blood." In other words, that the heart creates a vacuum during relaxation of the heartbeat.

(At least you didn't hold the common notion that the heart actively dilates because no lattice exists to pull the ventricles outward. Here's how you put it: "If there is no contrivance in the frame of the auricles or ventricles of the heart, by which they dilate themselves, I cannot conceive how they are dilated." And they don't, Sir; the heart muscle simply relaxes.)

You wisely rejected the proposition that fermentation in blood vessels causes pressure in the veins, which, in turn, dilates the heart upon return. You wrote that "the hearts of some animals continue to contract and dilate, or to beat, as 'tis commonly expressed, after they are separated from the other vessels and taken out of the Body." You correctly concluded, "If this be true, their dilatation is not caused by the force of the returning blood," but never offered a truly coherent counterproposal.

Unlike many of your colleagues, Dr. Franklin, you didn't became e   n-chanted with your own speculations: "I am not without apprehensions, that this hypothesis is either not new, or, if it is new, not good for any thing." For example, you erred on the cause of tingling when blood flow is cut off. To remind you: "let me add this particular. If you sit or lean long in such a manner as to compress the principal artery that supplies a limb with blood, so that it does not furnish a due quantity, you will be sensible of a pricking pain in the extremities like that of a thousand needles."

William Harvey

Your proposition "that the pricking pain is occasioned by the sides of the small vessels being pressed together" was incorrect. Obstructed blood flow in tiny vessels on the surface of nerves causes the tingling.

Regarding your many conjectures about bodily function, you displayed an admirably unpretentious stance when you told Colden, "If you give yourself the trouble of reading them, 'tis all I can modestly expect. Your silence about them afterwards will be sufficient to convince me, that I am in the wrong; and that I ought to study the sciences I dabble in, before I presume to set pen to paper."

You can't imagine, Dr. Franklin, how many people today put pen to paper without the slightest idea what they're writing about.

My next group of emails will cover those geologic processes that held your interest for so many years. For now, however, I'm taking on a challenging

Sudoku (a Japanese word meaning "single numbers") in the daily newspaper. Sudokus are simplified versions of what you called *Magical Squares*.

I know how much you enjoyed creating these objects. Recall writing this to Peter Collinson: "...in my younger days, having once some leisure, (which I still think I might have employed more usefully) I had amused myself in making these kind of magic squares and, at length, had acquired such a knack at it, that I could fill the cells of any magic square, of reasonable size, with a series of numbers as fast as I could write them, disposed in such a manner, as that the sums of every row, horizontal, perpendicular, or diagonal, should be equal; but not being satisfied with these, which I looked on as common and easy things, I had imposed on myself more difficult tasks, and succeeded in making other magic squares, with a variety of properties, and much more curious."

MAGICAL SQUARES.
Fig. 2.

200	217	232	249	8	25	40	57	72	89	104	121	136	153	168	181
58	39	26	7	250	231	218	199	186	167	154	135	122	103	90	71
198	219	230	251	6	27	38	59	70	91	102	123	134	155	166	187
60	37	28	5	252	229	220	197	188	165	156	133	124	101	92	69
201	216	233	248	9	24	41	56	73	88	105	120	137	152	169	184
55	42	23	10	247	234	215	202	183	170	151	138	119	106	87	74
203	214	235	246	11	22	43	54	75	86	107	118	139	150	171	182
53	44	21	12	245	236	213	204	181	172	149	140	117	108	85	76
205	212	237	244	13	20	45	52	77	84	109	116	141	148	173	180
51	46	19	14	243	238	211	206	179	174	147	142	115	110	83	78
207	210	239	242	15	18	47	50	79	82	111	114	143	146	175	178
49	48	17	16	241	240	209	208	177	176	145	144	113	112	81	80
196	221	228	253	4	29	36	61	68	93	100	125	132	157	164	189
62	35	30	3	254	227	222	195	190	163	158	131	126	99	94	67
194	223	226	255	2	31	34	63	66	95	98	127	130	159	162	191
64	33	32	1	256	225	224	193	192	161	160	129	128	97	96	65

Your most magical of magical squares

Within the last three years, puzzle-players have rediscovered magic squares, which are supplied in newspapers and books. Usually, a 9x9 format provides both numbered and empty squares; the puzzle-player must fill in the blanks. When your friend Jonathan Logan suggested that making magic squares might be trifling, you replied that it "may not be altogether useless, if it produces by practice an habitual readiness and exactness in mathematical disquisitions, which readiness may, on some occasions, be of real use." Indeed, Dr. Franklin, we now know that such puzzle work helps prevent senile dementia, explaining, perhaps, your own extended capacity.

GEOLOGY, EARTH SCIENCE & EVOLUTION

⊠ After the earthquake, an unusual
   redness appeared in the western sky   195

⊠ A purse made of the asbestos   199

⊠ One continent becomes old, another
   rises into youth and perfection   201

⊠ The surface of the globe would be a shell   205

⊠ With strata in various positions   211

⊠ The nature of this globe   215

⊠ The prolific nature of plants or animals   219

Hemispheres from Middleton's *Dictionary*

From: "Stuart Green" <stuartgreenmd@yahoo.com>
To: "Benjamin Franklin" <dr_benjamin_franklin@yahoo.com>
Subject: **After the earthquake, an unusual redness appeared in the western sky**

Dear Doctor Franklin:

I've often wondered how you felt when certain clergymen claimed that **you** caused Lisbon's great earthquake of 1755 and the one that rattled Boston a few weeks later. After all, earthquakes were seen as Divine retribution for a corrupt society, even when the heavenly justification for such widespread damage eluded understanding.

In ancient locations subjected to earthquakes, the god who shook the ground was the most feared—and the most revered. As with earthquakes, lightning bolts also served the objective of a punitive deity, although in a more precise manner. Zeus, Jupiter, Thor and Jehovah hurled them with great accuracy at those needing punishment.

The devil, as I mentioned previously, also had a hand in targeting lightning strikes, especially in medieval and Renaissance Europe. This seemed the only logical explanation for the frequent destruction by lightning of a town's highest and most godly structure, the church steeple.

During your era (now called the *Enlightenment*), scientifically minded individuals sought rational explanations for physical phenomena (although many natural philosophers stayed within bounds defined by the Holy Bible, especially the Genesis account of the world's origin).

Scientists of your generation viewed both lightning and earthquakes as explosions of flammable substances, either up in the clouds or below in caves. In the case of earthquakes, hollow regions of the planet, they believed, contained large quantities of sulfurous and nitrous substances that burned readily. Mine explosions confirmed the commonly held notion of combustible vapors within the earth.

According to your detractors, by directing lightning to the ground with "metallic points" you transferred destructive energy from sky to earth. Thomas Price, a Boston clergyman, offered his parishioners an explanation for the November 18, 1755 New England earthquake: Nitrous substances ignited in caves deep within the planet, causing the ground to shudder. While such destructive events occurred long before your invention, Price claimed the lightning rod—by directing fire from clouds into caves—increased the earth-shaking power of the event.

Fortunately for you, Sir, the accusations didn't last long, either in Europe or North America. A couple of days after Price's sermon, John Winthrop of Harvard came to your rescue, mocking Price's pernicious proposal.

Prior to a 1750 earthquake in London, English-speaking scientists had little interest in the subject. That temblor, however, certainly got their attention.

I assume it was a coincidence that your *Poor Richard's Almanack* for 1750 contained an account of a terrible earthquake that severely damaged the city of Port Royal, Jamaica, on June 7, 1692: "...the fine town was shaken to pieces, sunk into, and covered for the greater part, by the sea: By the falling of the houses, opening of the Earth, and inundation of the waters, nearly 2000 Persons were lost, many of note."

That report included a vivid description of the "harbour covered with the dead bodies of people of all conditions, floating up and down without burial." Moreover, the sea flowed through the cemetery and "washed the carcasses of those who had been buried out of their graves." You also mentioned that, "A sickness followed, [probably cholera] which carried off some thousands more." And finally, "during the earthquake, thieves robbed and plundered the sufferers, even among the ruins, while the Earth trembled under their feet."

(Some things never change, Dr. Franklin; whether earthquake, flood, or hurricane—robbers appear before rescuers.)

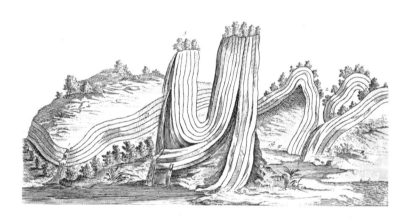

Geologic strata from Humphreys' *Nature Display'd*

Recall that you had previously written about earthquakes in the *Pennsylvania Gazette*. The first reference was to an odd occurrence in the early summer of 1732: "To the surprise of all the inhabitants on the Delaware, who live above Philadelphia, the water about a fortnight since, changed to a dark dirty red, so thick that 'tis said the fish could scarce see to get out of the way of boats, and were frequently struck by the oars."

"The conjectures of people are various concerning the cause of it," you reported, also mentioning a New York "flood that came down from the mountains, though they had had no rain, and overflowed the low lands, doing great damage."

Clearly you were speculating when you wrote, "an earthquake near the heads of both rivers, has forced out a quantity of subterraneous water into them." You cautiously reminded your readers, "These however are only conjectures. Time may possibly make us wiser."

Five years later, you reported that on Wednesday December 7, 1737, the "earthquake which surprised us here...was not felt at Annapolis in Maryland, but the accounts we have from New-Castle on Delaware, represent the shake to be nearly as violent there as here." There was a reason why "Three or four evenings successively after the earthquake, an unusual redness appeared in the western sky and southwards, continuing about an hour after sunset, gradually declining."

Your description of the red sunsets concluded with a discourse on the causes of earthquakes, copied without attribution from Chambers' 1728 *Cyclopedia*—a common practice in your times that wouldn't be tolerated today.

The article cited several theories about the origin of earthquakes, mostly variants of the concept that hollow portions of the earth contain explosive gases or, alternatively, that water or steam created by geothermal heat became plugged up, resulting in explosive back-pressure. Volcanoes erupted for the same reason, a remarkably accurate concept in light of today's knowledge of the events preceding a *pyroclastic* (major) eruption.

By the time of your announced demise, however, your own speculations about geologic processes has led at least one historian of science to characterize you as the greatest American geologist of the 18th century, a remarkable assessment, considering that you were neither a rock-hound nor a frequent visitor to geologic formations. Nevertheless, you did keep a small collection of interesting minerals to show visitors. I'll remind you of them in my next email.

How inclined strata appear horizontal
From Lyell's *Student's Elements of Geology*

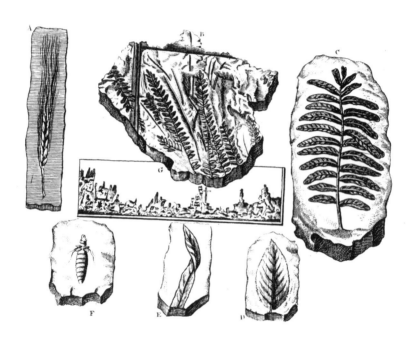

18th-century illustration of fossils from Middleton's *Dictionary*

From: "Stuart Green" <stuartgreenmd@yahoo.com>
To: "Benjamin Franklin" <dr_benjamin_franklin@yahoo.com>
Subject: **A purse made of the asbestos**

Dear Doctor Franklin:

Since we're on the subject of geology, I know of your collection of fossils, and your special interest in the large "elephant tooth" sent to you by George Croghan. You compared it to a typical elephant's tooth and noticed how it differed. After first mentioning that the tusks resembled those of African and Asiatic elephants, you told Croghan, "But the grinders differ, being full of knobs, like the grinders of a carnivorous animal; when those of the elephant, who eats only vegetables, are almost smooth." You also found it interesting "that elephants now inhabit naturally only hot countries where there is no winter, and yet these remains are found in a wintery country [Ohio]; and it is no uncommon thing to find elephants tusks in Siberia, in great quantities, when their rivers overflow and wash away the earth…"

We now know that Croghan had sent you a *Mastodon* tooth, belonging to an animal that once inhabited both North America and Russia but is now extinct. It turns out, Dr. Franklin, that animal and plant species appear and disappear on this planet through a process not understood until the grandson of your friend Erasmus Darwin worked out the mechanism. I'll tell you much more about this in a later email.

Speaking of rock hounds, I mentioned yesterday that you had a few mineral specimens you enjoyed showing guests. Do you recollect a visit by Swedish botanist Peter Kalm? He asked whether you had "come upon any evidence that places which were now part of the [North American] continent had formerly been covered with water." Based upon information from yourself and others, Kalm concluded that there were places in Philadelphia "which are at present fourteen feet or more under ground that were formerly at the bottom of the sea…"

When Kalm visited Philadelphia, you displayed three items the Swede found of interest, including a rock crystal—the largest he had ever seen. "It was four inches long and of a diameter of three finger's breadth," Kalm reported, adding, "I regretted it was not transparent but of a dingy watery color and opaque texture. All sides were smooth as if ground and had been found in Pennsylvania."

You also showed Kalm stalactites that "were discovered in a cave near Virginia." Kalm told Swedish readers that two kinds of such minerals existed, those hanging from ceilings and ones that "have been deposited like a round scraggy uneven fungus on the floor of it where the [calcareous] water has dripped from above."

Lastly, you demonstrated to Kalm a most remarkable mineral, asbestos. Kalm said it "was very dark gray mostly blackish and felt oily to the touch" and that fibers of the mineral can be made into paper. Indeed, Kalm told his readers that you had given him "some small pieces," which he kept.

We now know that asbestos fibers, when inhaled, can cause lung cancer. Before its health dangers became apparent, however, asbestos served as a fireproofing material for more than two centuries, so you correctly recognized the value of that fire-resistant mineral fiber.

As for asbestos, I was in London recently visiting the *British Museum*, a large collection of artifacts from around the world. It has a room that contains many instruments and publications from your era. On display are items from the collection of Sir Hans Sloane; I failed to spot the article you gave him

Hans Sloane

during your first visit to London, mentioned in your Autobiography: "I had brought over a few curiosities among which the principal was a purse made of the asbestos, which purifies by fire. Sir Hans Sloane heard of it, came to see me, and invited me to his house in Bloomsbury Square, where he showed me all his curiosities, and persuaded me to let him add that to the number, for which he paid me handsomely."

Another person who probably influenced your geologic thoughts, I suspect, was your one-time clerk, Lewis Evans. How well do you remember him?

After he left your employ, Evans wrote a book about the geology of British North America, published in 1755. The volume addressed seashells on mountaintops, leading Evans to conclude that geologic progression worked gradually over time, echoing ideas of French and German natural philosophers.

You, however, neither endorsed nor refuted his concept of a long time scale for geologic evolution. Nevertheless, your speculations about geology and earth science imply a stretched-out time frame, far beyond the biblical computations. More on this later.

From: "Stuart Green" <stuartgreenmd@yahoo.com>
To: "Benjamin Franklin" <dr_benjamin_franklin@yahoo.com>
Subject: **One continent becomes old, another rises into youth and perfection**

Dear Doctor Franklin:

Although your ideas about the formation of mountains and rivers and valleys followed accepted concepts of the time, you did come up with a unique proposal for the creation of continents that anteceded our modern understanding of *continental drift* and *plate tectonics* by more than 150 years.

In your writings, you occasionally mentioned trips to geologic wonders, caves, and rock formations. When you saw something intriguing, you often speculated on its origin. In some cases, your conjectures were ingenious. From August 8th to November 2nd, 1759, for example, while touring Scotland with your son William, you visited caverns at Derbyshire, followed by a descent into salt mines in Cheshire.

You wrote your brother Peter about sea-shells and bones and teeth of fish you found at high places and concluded that the mountains had either been under a deep sea that "has fallen away" or that the mountains were once much lower and have been "lifted out of the water to their present height by some internal mighty force." You correctly surmised that such a force could still be felt "when whole continents are moved by earthquakes."

Scientists of your era, Dr. Franklin, when proposing a Theory of the Earth, had to account for certain observations: first, cataclysmic events like earthquakes and volcanic eruptions; second, springs where water comes out of the earth hot and bubbling; third, geologic layers that were usually disrupted, folded or even vertically oriented in places; fourth, seashells and fish fossils embedded on mountaintops.

Other peculiar geophysical features—like movement of Earth's magnetic poles—also required explanation.

In response to gradual increasing geologic knowledge during the 17th century that seemed to refute the six-day biblical chronology, three English thinkers offered explanations faithful to established dogma.

Thomas Burnet concluded that a smooth featureless world existed before Noah's flood. Man's fall from grace led to events that disrupted the planet's features, tilted the earth's axis (creating seasons) and caused mountains to rise.

John Woodward, a 17th-century sage, proposed that geologic strata became exposed after the Flood to *help* mankind, who now must labor to produce sustenance.

William Whiston, in the late 17th century, agreed with Woodward about the benefit of terrestrial irregularities, but believed a passing comet caused the Great Flood.

You obviously knew of these Theories of the Earth; after all, you sold copies of Burnet's *Theory of the Earth* in your shop. Likewise, in your first *Poor Richard's Almanack* of 1733, you offered five computations of the Earth's age, including that of William Whiston. In later *Almanack* editions, you first ridiculed Whiston and then praised him as an authority on matters geological. However, as you learned more about Whiston's gloomy predictions of a looming world's end and comet theories, you abandoned 17th-century ideas about the earth's formation and brought yourself up to date.

The man responsible for the French translation of your book on electricity, the Comte de Buffon, in his *Natural History* contended that Earth and the other planets formed when a comet hit the sun at an oblique angle, knocking off some material.

Comte de Buffon

Buffon employed the idea that the Earth had an iron core—based upon the planet's magnetism—to compute the Earth's age. He heated iron balls and studied their cooling rates, concluding that the Earth was much older than Bishop Ussher's Bible-based estimation of 6,000 years.

Theologians faulted Buffon wherever he diverted from scriptural dogma. About thirty years after the first appearance of *Natural History*, Buffon produced an updated version: he concluded that the Earth was around 100,000 years old, with life first appearing on the planet after six lifeless epochs lasting 70,000 years.

By the time Buffon's revised edition appeared, you had been America's envoy to France for three years. I'd be surprised if you didn't discuss the Count's proposals with other philosophers and perhaps even with Buffon himself. Although you eventually developed a particularly ingenious theory of continental motion, your letters and other papers don't reveal any conjectures about the Earth's age or the speed of geologic processes.

James Hutton, whom you met on at least one trip to Scotland, is now regarded as the father of modern geology. His seemingly simple proposal proved such a threat to church doctrine that he felt compelled to search for additional evidence to bolster his claims.

James Hutton

Hutton's idea developed after he studied the same geologic region of Scotland for many years. He noted that rock formations eroded and turned to soil, but the process was extremely slow. The eroded soil, according to Hutton, washed into the sea where heat from the Earth's center fused the soil back into rocks again. Additionally, Hutton famously proposed that both basalt and granite, the seemingly solidest of earth's minerals, once flowed to the surface in a liquefied state.

Strata from Lyell's *Student's Elements of Geology*

More significantly, Hutton concluded that such geological processes have been progressing at slow rates for a long, long time—far longer than any biblical chronology would allow. The first printed exposition of Hutton's theory appeared in the Proceedings of the Royal Society of Edinburgh in 1788, two years before your *mort-faux*.

You never mentioned Hutton's ideas during the final years of your former life, busy as you were promoting abolition of slavery and enjoying your grandchildren. I've little doubt, however, that you'd have accepted Hutton's principal argument about the evolution and antiquity of geologic formations, because four years before Hutton made his ideas known, Dr. Franklin, you had concluded that, "One continent becomes old, another rises into youth and perfection."

Your subsequent correspondence on continent formation indicates, to my mind at least, that you gave the matter considerable thought.

I'll remind you of your geologic propositions in my next email. Now, however, I'm coming down with influenza, so I'll finish this geologic reminiscence later.

From: "Stuart Green" <stuartgreenmd@yahoo.com>
To: "Benjamin Franklin" <dr_benjamin_franklin@yahoo.com>
Subject: **The surface of the globe would be a shell**

Dear Doctor Franklin:

I'm feeling better now and will resume my communications about geology. (Remarkable though it may seem, we've not yet conquered influenza, although a vaccine is available. Each epidemic, however, involves a new strain of virus, so a fresh vaccine must be distributed to the populous.)

We call the modern unifying theory of the Earth *plate tectonics*; it properly accounts for the complimentary shapes of continents and many other phenomena as well. This hypothesis evolved during the last thirty years as scientists assembled odd bits of information derived from several sources. A Canadian, J. Tuzo Wilson, gets much credit for putting all the data together into a coherent theory of the Earth.

Before plate tectonics, however, there existed the much-disputed (but now confirmed) Theory of Continental Drift by Alfred Wegener, a German naturalist. And before that premise, you proposed your own remarkable conjecture of moving continents, as described in a 1782 letter to the French cleric, Abbé Jean Louis Giraud-Soulavie.

Unlike other philosophers who tried to explain the accumulating geologic knowledge of the time, Dr. Franklin, you offered Giraud-Soulavie a logical—though ultimately flawed—mechanism to raise mountains and move continents.

I don't know how familiar you were with Giraud-Soulavie's background. He spent much time hiking in the French mountains. There, he noted that geologic strata each contained distinct fossils embedded in the rocks. Giraud-Soulavie published a series of seven volumes describing his findings about the scale of geologic time. He had to withdraw and re-issue the first book (1780) because it contained conclusions that didn't sit well with his church's authorities.

Giraud-Soulavie realized that layers of rocks contained a succession of fossils with increasing shell complexity in the top layers. He noted that the deepest layer contained shells of primitive creatures with no living examples. The next layer up held both extinct forms and others resembling living animals. This pattern followed up to the topmost level, which contained fossilized remains of present-day plants and animals.

This discovery suggested a rather old age for the Earth, but more importantly, pointed towards using fossils within different layers to compare one region's geologic formations to those elsewhere. The principle, which assumed that identical creatures inhabited different locations at the same time, enabled *stratigraphic analysis* of rock formations.

Perhaps to please his ecclesiastical superiors, Giraud-Soulavie inserted at least one catastrophe into the geologic sequence, thereby bringing his proposal in line, more or less, with the Genesis account of Creation. I'm not sure why, but you contacted Giraud-Soulavie in 1781 and asked the cleric for proof that a catastrophic event occurred, perhaps because you had considered such a possibility while you visited the caves in Derbyshire. After additional correspondence and possibly a face-to-face meeting or two, you wrote to Giraud-Soulavie about your Theory of the Earth.

That manuscript, now in the Historical Society of Pennsylvania's collection, contains enough original ideas to insure you a place in the history of science even if you had never done anything else of note. Your letter to Giraud-Soulavie was presented publicly two years before your 1790 encaskment at a meeting of the American Philosophical Society; five years thereafter it was published in their *Proceedings*. As a result, the document influenced geologists for most of the 19th century, until your core hypothesis [pun intended] fell by the wayside.

To remind you, Dr. Franklin, you started by noting that oyster shells within stone at the lowest exposed part of the "calcerous rock" [chalk cliffs] in Derbyshire implied that a great upheaval had occurred to cause "some part of it having been depressed under the sea, and other parts which had been under it, being raised above it." You rightly concluded that such movement of superficial landmasses was "unlikely to happen if the Earth were solid to the centre."

You thus logically "imagined that the internal part might be a fluid more dense and of greater specific gravity than any of the solids we are acquainted with, which therefore might swim in or upon that fluid."

Your conjecture here was absolutely correct: The Earth's outermost features, including the continents, do indeed swim on a fluid that is much denser than the surface structures.

You appropriately took your idea one step further: "Thus the surface of the globe would be a shell, capable of being broken and disordered by any violent movements of the fluid on which it rested." You erred, however, when you imagined the fluid in the planet's center to be compressed air, made so dense by gravitational pressure that water would float on its surface. (Here's how you put it: "And as air has been compressed…to be twice as dense as water, in which case…the air would be seen to take the lowest place, and the water to float above and upon it.")

The air, in your conjecture, being compressed by gravitational forces, "would at the depth of leagues be heavier than gold." At the planet's center, you, like others of your time, surmised the existence of fire—the source of heat for hot springs and volcanoes. The compressed air surrounding the central fire, because of its density, would expand with enough force to "move the surface" when heated by the central fire.

You also suggested that the compressed air could even "be of use in keeping alive the central fires." Steam from water coming in contact with the central fire might also "be an agent sufficiently strong" to move the "incumbent Earth."

You followed Isaac Newton's notion of God as the author of the laws of physics that, in turn, control nature and the world. "If one might indulge imagination in supposing how such a Globe was formed," you said, all the elements were separate particles in the beginning; they were "mixed in confusion and occupying a great space."

In your own version of Genesis, you imagined that God first "ordained gravity or the mutual attraction of certain parts, and the mutual repulsion of other parts," that got the process started. Various particles would coalesce or separate, according to their mutual attraction or repulsion. Material would arrive at regions with the "same specific gravity with themselves," and would stay at that level. This process "would form the shell on the first Earth."

At the shell stage, you had concocted a stable Earth, with the elements layering out according to specific gravity and the effect of attraction and repulsion. You needed, however, a motive force to stir things up a bit, so you cleverly created two of them—the "movement of the parts towards their common centre, could naturally form a whirl," and "the turning of the new formed globe upon its axis."

You obviously knew that careful measurements revealed that the Earth was slightly larger around the equator than around the poles, a consequence of our planet's rotation. You thus envisioned the newly formed Earth as a slightly flattened sphere—bulging at the equator—with a hot gaseous center surrounded by a layer of air that was compressed by gravitational forces to the density of gold at its deepest level (yet perhaps capable of sustaining the fire within).

The air further from the center, if I correctly understand your view, was less dense because it experienced less compression. Water and solid matter floated within or on the compressed air at each substance's neutral buoyancy level. The surface of your new planet would, of necessity, be a smooth shell since the laws of physics caused the elements of differing densities to layer-out.

By starting out with a smooth-shelled planet, Dr. Franklin, you borrowed ideas from the British philosophers of the late 1700s (Burnet, Woodward and Whiston), yet you never assumed a divinely ordained deluge or passing comet to wrinkle the surface. Unless you came up with a mechanism to disturb the newly formed Earth's equilibrium, your neo-planet would remain smooth-surfaced forever.

Your solution to the problem, although incorrect, was clever indeed: A change in the planet's axis of rotation, even a slight one, would be enough to break apart the surface because the outer shell's equatorial bulge would no longer be in the correct position. Recall how you put it: "If by any accident afterwards the axis should be changed the dense internal fluid by altering its form must burst the shell and throw all its substance into the confusion in which we find it."

You next came up with a mechanism to change the planet's axis: the gradual cooling of molten iron. You first reminded Giraud-Soulavie about his own observation "on the ferruginous [iron] nature of the lava which is thrown out from the depths of our volcanoes." This meant that liquid iron existed below the Earth's surface. "It has long been a supposition of mine," you wrote, "that the iron contained in the substance of this globe, has made it capable of becoming, as it is, a great magnet." You also assumed that "magnetism exists perhaps in all space; so that there is a magnetical north and south of the universe as well as of this globe."

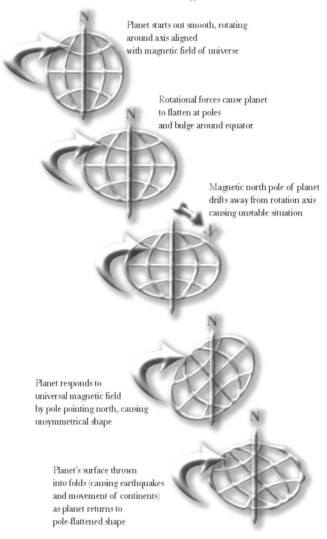

Planet starts out smooth, rotating around axis aligned with magnetic field of universe

Rotational forces cause planet to flatten at poles and bulge around equator

Magnetic north pole of planet drifts away from rotation axis causing unstable situation

Planet responds to universal magnetic field by pole pointing north, causing unsymmetrical shape

Planet's surface thrown into folds (causing earthquakes and movement of continents) as planet returns to pole-flattened shape

Franklin hypothesis of the creation of continents, earthquakes and other geologic processes

Such a universal magnetic field had value for future travel to outer space, you said: "if it were possible for a man to fly from star to star, he might govern his course by the compass."

The universe's magnetic field, in your conjecture, would impart magnetism to the molten iron in our newly formed planet. Because the hot iron is "soft," its magnetism would be "naturally diffused equally throughout." This magnetized molten iron would change shape as it flowed within the universe's magnetic field. "If it cools or grows hard in that situation," you claimed, "it becomes a permanent one" [*i.e.*, a magnet].

Because the cooled iron differs in shape from the earlier molten iron, the Earth's axis would change as the magnetized iron re-aligned itself with the universe's magnetic field. The planet's axis-shift "occasioned the rupture of its shell, the submersion and emersions of its lands and the confusion of its seasons."

To account for *future* geophysical changes, you suggested that the Earth's axis may gradually shift as much as 90 degrees to "place it in the present equator and make the new equator pass thro' the present poles." When this happens, the planet's shape will change: the present equatorial bulge will flatten and the current polar flattening will protrude. "It is easy to conceive," you wrote, "what a sinking of the waters would happen in the present equatorial regions, and what a rising in the present polar regions, so that vast tracts would be discovered that now are under water, and others covered that now are dry, the water rising and sinking in the different extremes near five leagues" [15 miles].

Since you seemed unsure about the possibility of future axis changes, you stated, "The globe being now become a permanent magnet we are perhaps safe from any future change of its axis." The situation, however, wasn't entirely stable; in your conjecture, forces below the surface remained active.

You rightly assumed the existence of an "internal ponderous fluid" that, by its movement, could cause both earthquakes and volcanoes. Such movement might occur, you surmised, when subterranean water came in contact with the internal "fire under the Earth," causing an explosion.

The underground blast, another notion you borrowed from predecessors, "not only lifts the incumbent earth that is over the explosion, but impressing with the same force the fluid under it, creates a wave that may run a thousand leagues, lifting and thereby shaking successively all the countries under which it passes."

Your letter to Giraud-Soulavie offered a Theory of the Earth that explained many geophysical observations: Earth's seemingly hot core; the planet's shape; shells of undersea creatures on mountaintops; molten iron spewing from volcanoes; and earthquakes.

You didn't, however, explain the layering of rock formations so easily accounted for by the sedimentary hypotheses of Hutton and other geologists. I know you were aware of theories about how geologic strata developed and even contemplated your own ideas on the subject, which you didn't include in the Giraud-Soulavie letter. I'll say more about this topic in my next email.

Alfred Wegener

J. Tuzo Wilson

From: "Stuart Green" <stuartgreenmd@yahoo.com>
To: "Benjamin Franklin" <dr_benjamin_franklin@yahoo.com>
Subject: **With strata in various positions**

Dear Doctor Franklin:

Recall that when you went to London for the second time in 1757, you became acquainted with John Michell, an English astronomer and geologist. (Michell is remembered by posterity for the Michell-Cavendish experiment, a method of determining the Earth's mass with a simple piece of laboratory equipment and Newton's gravitation formulae.)

Five years after the 1755 Lisbon earthquake, Michell wrote an article for The Royal Society's *Philosophical Transactions* titled "Conjectures Concerning the Cause, and Observations upon the Phaenomena of Earthquakes." Before submitting the work for publication, however, Michell sent it to your mutual friend John Pringle who in turn passed the piece along to you for perusal.

Michell, in his manuscript, linked volcanoes and earthquakes together in a hypothesis attributing both events to subterranean steam pressure. The steam, he proposed, arose when underground water came in contact with the Earth's hot interior. The explosive force of either a volcanic eruption or an earthquake disordered the previously regular strata, causing the jumbled layers we see today.

You thanked Pringle for sending the article; it contained ideas you later included in the Giraud-Soulavie letter.

Always practical, I believe you saw beyond the destructive nature of earthquakes and volcanoes and looked to their benefit. You told Pringle, "Had the different strata of clay, gravel, marble, coals, lime-stone, sand, minerals &c. continued to lie level, one under the other, as they may be supposed to have done before these convulsions, we should have only the use of a few of the uppermost of the strata, the others lying to deep and too difficult to come at."

The Earth's occasional convulsions, by throwing up the strata into oblique positions, had brought "a great variety of useful materials" to the surface, "which would otherwise have remained eternally concealed from us." Our planet's paroxysms, rather than being a "ruin suffered by this part of the universe," were actually a means of making Earth "more fit for use, more capable of being to mankind a convenient and comfortable habitation."

After considering your suggestions, it appears that Michell modified his original manuscript to include certain gradualist ideas he obtained from you but which originally came from your former assistant, Lewis Evans.

Your final written foray into geological speculation was contained in a letter to James Bowdoin, the Massachusetts merchant. Bowdoin, in early May 1788, informed you that New Hampshire would likely ratify the new U.S. Constitution, thereby making the document the official governing instrument of those states agreeing to abide by it.

I was amused that your response to Bowdoin, rather than dwelling on political issues, got back to the subject of your "ancient correspondence," namely, speculations about science.

You told Bowdoin that your mutual friend John Winthrop (who defended you against the earthquake-provoking charges of Reverend Thomas Price) "made me the compliment that I was good at starting games for philosophers. Let me try if I can start a little for you." You then posed a number of Socratic questions about earth science for Bowdoin to think about. Many of the queries relate back to ideas contained in your Giraud-Soulavie letter six years earlier.

For the most part, your questions dealt with Earth's magnetism. You asked, "Is it likely that iron ore immediately existed when this globe was first formed, or may it not rather be supposed a gradual production of time?" In other words, has the amount of iron ore changed with time, or was the current amount present from the very beginning?

Next you asked Bowdoin (rhetorically, of course) if there had been a time when our planet didn't have magnetic polarity. Repeating a theme from the Giraud-Soulavie letter, you wondered whether the Earth had received its magnetic polarity from an external source, specifically, the rest of the solar system. Once again, you raised the issue of future space travel: "In short, may not a magnetic power exist throughout our system, perhaps thro' all systems, so that if men could make a voyage in the starry regions, a compass might be of use?" Moreover, would not such a universal magnetic field "be serviceable in keeping the diurnal revolution of a planet more steady to the same axis?"

For some inexplicable reason you diverted from your Giraud-Soulavie letter's hypotheses by adding an idea from Whiston's passing comet theory: "As the poles of magnets may be changed by the presence of stronger magnets, might not in ancient times the near passing of some large comet, of greater magnetic power than this globe of ours, have been a means of changing its poles, and thereby wracking and deranging its surface, placing in different regions the effect of centrifugal force, so as to raise the waters of the sea in some while they were depressed in others?"

Clearly, you approved the idea of magnetic pole migration as the motive force of geophysical change and weren't about to let go: "Is not the finding of great quantities of shells and bones of animals (natural to hot climates) in the cold ones of our present world, some proof that its poles have been changed?"

You even took the notion of a past great flood as a fact of geologic history and asked Bowdoin: "Is not the supposition that the poles have been changed, the easiest way of accounting for the Deluge, by getting rid of the old difficulty how to dispose of its waters after it was over" by changing the Earth's shape?

Without mentioning the difference between the equatorial and transpolar diameters of the planet, you asserted that the seas would fall "about 15 miles in height, and rise as much in the present polar regions" if the Earth's rotational axis changed and the poles were "placed in the present equator." You then added an idea not mentioned in the Giraud-Soulavie letter: "the effect would be proportionable if the new poles were placed any where between the present [position] and the Equator."

As with Giraud-Soulavie, you proposed that the planet's surface features floated on a super-dense substance: "Does not the apparent wrack of the surface of this globe, thrown up into long ridges of mountains, with strata in various positions, make it probable, that its internal mass is a fluid, but a fluid so dense as to float the heaviest of our substances?" You again suggested that if air's density increased towards the Earth's center in the same proportion that it does on the surface, "at what depth may it be equal in density with gold?"

"Can we easily conceive," you asked, "how the strata of the Earth could have been so deranged, if it had not been a mere shell supported by a heavier fluid? Would not such a supposed internal fluid globe be immediately sensible of a change in the situation of the Earth's axis, alter its form and thereby burst the shell, and throw up parts of it above the rest?"

Chasm formed by the earthquake of 1783, near Oppido, in Calabria
From Lyell's *Principles of Geology*

Finally, bringing earthquakes into your Theory of the Earth, you asked, "Might not a wave by any means raised in this supposed internal ocean of extremely dense fluid, raise in some degree as it passes the present shell of incumbent Earth, and break it in some places, as in earthquakes?" As the planet's surface broke up, the "progress of such waves" could account for the "the rumbling sound being first heard at a distance, augmenting as it approaches, and gradually dying away as it proceeds" during an earth-quake. Here you quoted "a very ingenious Peruvian" you met in Paris who told you about such a sound that traveled from north of Lima all the way south "quite down to Buenos Ayres" during their last great quake.

The Bowdoin letter proved your last known communication on the sub-ject of earth science. Like the Giraud-Soulavie letter, it was read before the American Philosophical Society and established you as a theoretical geolo-gist before field research rose to such prominence in the discipline.

In tomorrow's email I'll tell you how close you came to our present-day understanding of geologic process, as insightful geologists, especially Charles Lyell, promoted Hutton's gradualism. You'll be amazed.

Charles Lyell

From: "Stuart Green" <stuartgreenmd@yahoo.com>
To: "Benjamin Franklin" <dr_benjamin_franklin@yahoo.com>
Subject: **The nature of this globe**

Dear Doctor Franklin:

The principle issue facing geologists after your death came from the accumulating evidence that the planet was far older than had been proposed by most 18th-century natural philosophers. Interestingly, you never joined that particular debate. Although your axis-changing, fire-and-air cored Earth must have taken quite some time to transition from a smooth planet to one with its present surface irregularities, you didn't, as far as I know, put a precise time scale on the process.

Geologists have now concluded that the earliest known rocks on the planet are 3.96 billion years old.

The debate about the makeup of Earth's core—gas, liquid or solid—heated up through the 19th century. Geologists resolved the issue by studying how seismic waves travel through the planet after an earthquake or man-made super-explosion. You, Sir, by assuming that seismic waves propagate through a sub-surface liquefied matter, proved farsighted once again.

Recall telling Oliver Neave about the way sound waves can travel further through certain substances—wood or water—than through air. Here's how you put it: "It is a well-known experiment, that the scratching of a pin at one end of a long piece of timber, may be heard by an ear applied near the other end, though it could not be heard at the same distance through the air. And two stones being struck smartly together under water, the stroke may be heard at a greater distance by an ear placed underwater in the same river, than it can be heard through the air."

The same thing happens with seismic waves. They traverse different layers of the planet at different speeds depending on the material's composition. And, as with sound waves, they reflect off surfaces between layers deep in the Earth.

Geologists, using such waves and other measurements, have determined that our planet's core is 4320 miles in diameter and has two components: a solid inner core (1520 miles in diameter) and a liquefied outer core (1400 miles thick). The inner core is composed primarily of iron alloyed with other elements and has a density equal to that of 18-carat gold. (Yes, Dr. Franklin, I do recall your conjecture that air at the planet's center was compressed by gravity to the density of gold.) The outer core is mostly an iron-nickel alloy and has the density of 9-carat gold. As you and others of your era surmised, the core is hot as Hades: 10,000°F in the center, 8,000°F at the edge.

Surrounding the core is the 1800-mile thick mantle made of slow flowing minerals. Floating on the mantle's upper portion is the hard *lithosphere*—your conjectured shell.

The outermost rocky *crust* sits atop the lithosphere, varying in thickness from 4 miles (beneath the oceans) to 25 miles (on the continents).

The structure of the Earth

Your theory from 250 years ago attributed crustal motion of the Earth's shell to a centrifugally created change in planetary shape. Your motive force: an axis-shift of the magnet poles. (It is, in fact, our planet's gradual cooling that energizes the process.)

The modern theory of plate tectonics assumes the presence of about 20 plates of various sizes forming the Earth's surface. They move slowly with respect to each other, with the thinner ones usually sliding under the thicker ones, lifting mountains and causing earthquakes. (The continents are firmly affixed to the thickest plates.) As the plates slide under one another, they curl downward and melt into the slow-moving liquefied mineral layer that constitutes the Earth's mantle. There, the material circulates into deeper layers and then re-emerges into the lithosphere once again at the place of *sea-floor spreading*, thus accounting for the movement of continents.

Because the theory of plate tectonics explains the basis for earthquakes, volcanoes, mountain building and numerous other features of our planet, the concept is central to modern geology.

The Earth's tectonic plates are remarkably thin when compared to the overall size of the planet, constituting less then 0.7 % of the global radius. In fact, when the Earth is shown in diagrammatic cross-section, the planet resembles an egg, with a hard but brittle-looking shell and a mostly liquid center.

Keeping this image in mind, Sir, recall how you explained the consequences of your axis shifting theory to James Bowdoin in 1788: "if we could alter the position of the fluid contained in the shell of an egg, and place its longest diameter where the shortest now is, the shell must break; but would be much harder to break if the whole internal substance were as solid and hard as the shell?"

You, more than any philosopher of the 18th century, recognized the speculative nature of your conjectures, telling Giraud-Soulavie, "Superior Beings smile at our theories, and at our presumption in making them." You, however, saw value in theory making: "If they occasion any new enquiries and produce a better hypothesis, they will not be quite useless."

Wisely, you also recognized that your speculations about nature weren't nearly as valuable as the field research Giraud-Soulavie did in France's mountains: "You see I have given a loose to imagination; but I approve much more your method of philosophizing, which proceeds upon actual observation, makes a collection of facts and concludes no farther than those facts will warrant." You ended the Giraud-Soulavie letter by saying, "In my present circumstances, that mode of studying the nature of this globe is out of my power, and therefore I have permitted myself to wonder a little in the wilds of fancy."

What so astonishes me, Sir, is how you kept up your correspondence with fellow scientists in spite of momentous events of the time. Considering the historic consequences of your war-ending negotiations with the British, America today is grateful, Dr. Franklin, that you didn't do extensive field research at geologic formations.

Once Hutton and Lyell established gradualism in geologic processes, the relationship between fossils in successive layers became apparent, with more complex plants and animal fossils embedded in higher, more recent, strata and simpler forms in lower, earlier levels. Clearly, life evolved over time. I'm sure you'll enjoy reading about the discovery of nature's method of species evolution because you played an early, if minor role in the story, which featured Charles Darwin, scion of a family well known to you, and a man of the cloth named Robert Malthus.

Robert Malthus

Charles Darwin

From: "Stuart Green" <stuartgreenmd@yahoo.com>
To: "Benjamin Franklin" <dr_benjamin_franklin@yahoo.com>
Subject: **The prolific nature of plants or animals**

Dear Doctor Franklin:

Your friends Erasmus Darwin and Josiah Wedgwood were the two grandfathers of a future English naturalist, Charles Darwin. In like manner, your essay on population growth grandfathered Charles Darwin's most famous book, *Origin of the Species*.

You and Erasmus seemed kindred spirits, both having developed phonetic alphabets to replace the peculiar spellings of English. Ten years after your alleged death, Erasmus Darwin wrote a book proposing that all species, rather than being forever fixed, changed over time. In his *Zoonomia or, the Laws of Organic Life,* Erasmus claimed that living creatures somehow willed certain bodily changes for their own best interest and that these alterations passed to subsequent generations. Moreover, he raised this interesting query: "Can it be that one form of organism has developed from another; that different species are really but modified descendants of one parent stock?"

Charles Darwin, who eventually answered the question posed by Erasmus, didn't learn about the mysteries of evolution while sitting on his grandpa's lap, because Erasmus died seven years before Charles was born. Perhaps Robert Darwin, son of Erasmus and father of Charles, transmitted to the future naturalist his grandfather's curiosity, so necessary for solving the puzzle of species formation.

Do you remember meeting 21-year-old Robert Darwin in Paris shortly before you left for America? I ask because Erasmus, in 1787, sent you a letter thanking you for the encounter. By that time, Erasmus Darwin, like many of his countrymen, believed that you almost single-handedly created the United States of America. I image you felt particularly proud when Erasmus proclaimed, "I can scarcely forget that I am also writing to the greatest statesman of the present, or perhaps of any century who spread the happy contagion of liberty among his countrymen; and like the greatest man of all antiquity, the leader of the Jews, delivered them from the house of bondage, and the scourge of oppression."

Robert Darwin became a physician like his father and hoped his son Charles would do the same, but the youngster had other ideas. After completing his studies at Cambridge, Charles accepted an opportunity to travel as a naturalist on a far-ranging sea voyage.

The five-year journey provided Charles Darwin plenty of time to read and to gather specimens of plant and animal life for future study. He was especially intrigued by *Principles of Geology*, written in 1830 by British geologist Charles Lyell, who claimed that geological processes usually occur over great time spans, rather than suddenly as with catastrophic floods and the like. (You clearly had an interest in this subject, which I've already reviewed for you.)

Charles Darwin recognized that this gradualism approach might explain the origin of the new species, which seemed to have replaced the fossilized ones he found during his voyage. He rejected his grandfather's notion that animals could will changes in themselves and successive generations.

Upon return to England, Charles joined a pigeon fancier's club to better understand how breeders gradually modify the shape of animals. Nest mates and littermates differ ever so slightly in numerous characteristics. By mating a pair of animals with a preferred trait, breeders ensure transmission to the next generation.

How, Charles Darwin wondered, could species change in nature, where no breeder intervenes to select features for progressive amplification?

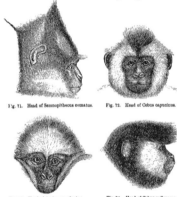

Fig. 71. Head of Semnopithecus comatus. Fig. 72. Head of Cebus capucinus.

Fig. 73. Head of Ateles marginatus. Fig. 74. Head of Cebus vellerosus.

From Darwin's *The Descent of Man*

In his *Autobiography* Charles described how he stumbled upon the answer: "In October 1838, that is, fifteen months after I had begun my systematic inquiry, I happened to read for amusement Malthus on *Population*, and being well prepared to appreciate the struggle for existence which everywhere goes on from long-continued observation of the habits of animals and plants, it at once struck me that under these circumstances favourable variations would tend to be preserved, and unfavourable ones to be destroyed. The results of this would be the formation of a new species. Here, then I had at last got a theory by which to work."

Thus did Charles Darwin realize that *overpopulation* and *competition for resources* such as food act as the breeder, selecting those best suited to survive and to produce descendents with the more useful traits. In the notebook where Darwin recorded thoughts about his research, the entry for September 28, 1838 reads, "the warring of the species as inference from Malthus."

Charles Darwin never failed to credit Thomas Malthus with the idea of overpopulation and competition. He wrote, "This is the doctrine of Malthus, applied to the whole animal and vegetable kingdoms. As many more individuals of each species are born than can possibly survive...there is a frequently recurring struggle for existence...any being, if it vary however slightly in any manner profitable to itself...will have a better chance of surviving, and thus be *naturally selected*. From the strong principle of inheritance, any selected variety will tend to propagate its new and modified form."

So who was Malthus, and what did he say that so impressed Charles Darwin?

Thomas Robert Malthus was born on February 17, 1766, four days after you famously testified before the House of Commons to help get the Stamp Act repealed. Indeed, if serendipity plays any role in the affairs of mankind, Malthus dropped within the womb at the exact moment you, responding to the question, "In what proportion hath population increased in North America?" gave the Malthusian reply: "I think the inhabitants of all the provinces together, taken at a medium, double in about 25 years."

Malthus, a local country parson, might have ministered his flock in perpetual obscurity but for a book called *The Enquirer* written by another clergyman, William Godwin. Malthus disagreed with the book's premise that offered a positive vision of mankind's future. Godwin's optimism matched that of your friend, the French mathematician Marquis de Condorcet, who presented this hopeful projection of mankind's future: Self-serving monarchs and priests would be abolished, religion debunked, and inequality between social classes and nations eliminated.

Malthus, in response, proposed a much gloomier outlook. In 1789 he anonymously published a pamphlet entitled *An Essay on the Principle of Population, as it Affects the Future Improvement of Society with Remarks on the Speculations of Mr. Godwin, M. Condorcet, and Other Writers.* Malthus claimed that populations grow faster than the food supply, causing diseases due to overcrowding and poor nutrition, as well as wars for resources. He concluded that only by limiting the number of children in a family could we prevent catastrophic competition for resources.

Malthus didn't cite any authorities to support his conclusions, although he made good use of birth and death records. He wrote an enlarged edition three years after the first—this time with his name proudly displayed on the front page. He shortened the title to *An Essay on the Principle of Population,* and saw the work through multiple editions, the final one written in 1806.

In all editions after his first anonymous one, Malthus rightfully acknowledged the distinct contribution you made to population theory. In fact, your name appears seven times in *An Essay on the Principle of Population.* At the beginning of the *Essay,* for instance, Malthus wrote, "It is observed by Dr. Franklin that there is no bound to the prolific nature of plants or animals but what is made by their crowding and interfering with each other's means of subsistence. Were the face of the earth, he says, vacant of other plants, it might be gradually sowed and overspread with one kind only, as for instance with fennel: and were it empty of other inhabitants, it might in a few ages be replenished from one nation only, as for instance with Englishmen."

Malthus was quoting, of course, from your seminal 1751 essay *Observations concerning the Increase of Mankind and the Peopling of Countries.*

I realize you wrote *Observations* with two goals: first, to persuade the British to take as much of North America from the French as possible; second, to convince England that allowing more manufacturing in colonial North America wouldn't harm industry in the Mother Country. Nevertheless, according to Malthus' modern biographer, when Malthus came across your last (most comprehensive), edition of *Experiments and Observations*, he "must have felt, as he read it, that almost all his ideas and discoveries had been compressed into this one exquisite nutshell years before he himself was born."

Considering how well known were *Observations*, one wonders: Did Malthus actual encounter your ideas *before* he wrote the first edition of his *Essay on Population*? While it seems I'm humoring you by ascribing significance to the role your writings had on Malthus, this proposal has been raised before. Karl Marx, a German philosopher and social historian, came to the same conclusion.

Being a utopian, Marx disagreed with the gloomy predictions of Thomas Malthus. For this and other reasons, Marx wrote in his 1867 *Das Kapital* that the *Essay on Population* by Malthus "is nothing more than a schoolboyish superficial plagiary of De Foe, Sir James Stewart, Townsend, Franklin, Wallace, &c., and doesn't contain a single sentence thought out by himself..."

Marx marveled at your analysis of society's economic forces because, in the beginning of *Das Kapital*, he wrote, "The celebrated Franklin, one of the first economists... who saw through the nature of value, says: 'Trade in general being nothing else but the exchange of labour for labour, the value of all things is...most justly measured by labour.' "

Marx contended that you erred, however, by not distinguishing one form of labor from another and thus "reduces them all to equal human labour...without further qualification, as the substance of the value of everything."

Also, Marx supported his claim of manual work's value—in conjunction with tool use—by saying that, "Franklin therefore defines man as a tool-making animal."

Elsewhere in *Das Kapital*, Marx buttressed his contention that a merchant "parasitically shoves himself in between" producer and buyer by quoting your statement that "War is *Robbery*, Commerce is generally *Cheating*" (I wonder what you, a champion of business enterprise, will say about reading your ideas in *Das Kapital*, a book encompassing Marx's utopian vision that supported his call for a workers' revolution.)

By writing a significant tract on population dynamics—now called *demographics*—that emphasized the relationship of birthrate, overpopulation, competition for resources and reproductive success, you provided useful support to the population theory, which Malthus based on all these factors.

Reading Malthus, in turn, gave Charles Darwin the insight he needed to formulate his historic proposals that have profoundly impacted our species in many different ways.

ATMOSPHERIC SCIENCE & TECHNOLOGY

✉ *Suppositions and Conjectures on the Aurora Borealis*   225

✉ The balloon we now inhabit   227

✉ Thermometers are often badly made   233

✉ The dissolved water in the air   237

✉ As easy as pissing abed   239

✉ Whirlwinds at land, waterspouts at sea   243

✉ Some curious and useful discoveries in meteorology   245

✉ Different degrees of heat   249

✉ The author's description of his Pennsylvania fire-place   251

✉ I soon became a thorough Deist   253

✉ A fool and a madman   255

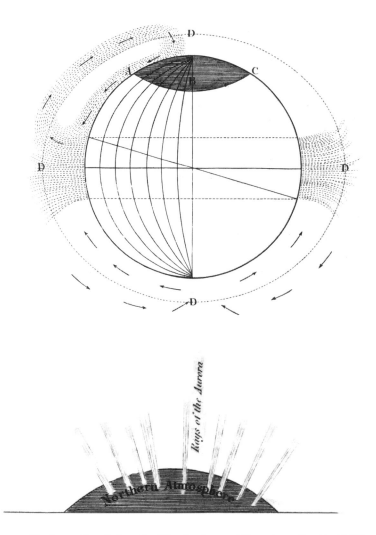

The Aurora Borealis from Franklin: *Experiments and Observation on Electricity (4th Ed.)*

From: "Stuart Green" <stuartgreenmd@yahoo.com>
To: "Benjamin Franklin" <dr_benjamin_franklin@yahoo.com>
Subject: ***Suppositions and Conjectures on the Aurora Borealis***

Dear Doctor Franklin:

I'll start this section with the Aurora Borealis, now also called *The Northern Lights*. I regret to inform you that your conjecture about that extraordinary spectacle has proven wrong, although not by much. I'd be remiss, however, if I didn't tell how much I marvel at your draft document called *Suppositions and Conjectures on the Aurora Borealis*. It's a complete theory; you started by describing basic physical principles of heat and cold, how hot air rises and cold air, being heavier, descends.

You next reminded readers that rain and clouds are electrified, an observation that accounts for lightning. The Earth's poles, being colder than the equator, experience a downward flow of air, according to your hypothesis, causing the warm tropic air and its electricity-laden clouds to move towards the polar regions. You wrote, "The great cake of ice that eternally covers those regions may be too hard frozen to permit the electricity descending with that snow to enter the Earth. It will therefore be accumulated on that ice."

Aurora Borealis with Big Dipper in center (my own image)

You reinforced your argument by stating that the electricity, which cannot penetrate the ice to reach the ground, acts as though in a "bottle overcharged," so it flows along the atmosphere towards the equator. You next asked the rhetorical question, "If such an operation of nature were really performed, would it not give all the appearances of an *Aurora Borealis?*" Perhaps it might, Dr. Franklin, but that isn't how the Northern Lights form.

The Aurora Borealis occurs when charged particles, mostly electrons, collide with gases in the earth's outer atmosphere. Recall that Earth is a giant magnet. Like any magnet, lines of magnetic force emanate from the north and south poles. Electrons flow in spiral pathways along those lines. When the spiraling electrons strike atoms in the upper atmosphere, light is emitted.

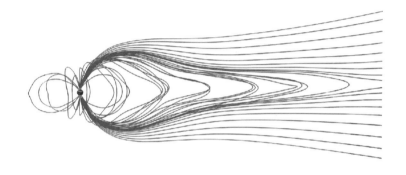

Lines of magnetic force surrounding the earth
The sun is to the left.
Image from NASA

Thus, in a sense, your conjecture about the Aurora Borealis, although not correct, contained certain elements of truth. For one thing, an electrical phenomenon accounts for the illumination. Moreover, the Earth's physical properties explain the location of the lights near the poles.

Considering that you made your proposal during the 18th century, when many philosophers clung to ancient ideas and superstitions, I congratulate you for trying to logically account for an observed phenomenon without resorting to supernatural explanations.

From: "Stuart Green" <stuartgreenmd@yahoo.com>
To: "Benjamin Franklin" <dr_benjamin_franklin@yahoo.com>
Subject: **The balloon we now inhabit**

Dear Doctor Franklin:

Twice a day throughout the year, *meteorologists* (weathermen) around the world simultaneously release 1100 lighter-than-air atmospheric balloons. Measuring 6 feet in diameter, the balloons expand to many times that size as they lift into the air, each carrying aloft a small instrument package that measures temperature, humidity and barometric pressure and transmits that information back to earth electrically. The balloons, at about twenty miles up, burst apart and the instrument assemblies drift back to the planet's surface with special kites called *parachutes*.

You, Sir, would relish each weather balloon launch because you held such a resolute interest in both ballooning and atmospheric science.

Balloons, you'll surely remember, evolved in France during the time you were finalizing the treaty that ended America's War of Independence. With all you did during the negotiations, it's remarkable that you had any time for balloons, but involved you were, Dr. Franklin—as a spectator, reporter, and eventually, the source for non-French enthusiasts wanting information about the exciting new technology.

Recall that the first public balloon demonstration occurred on June 5, 1783, at Annonay, a city in the south east of France. The contraption was the brainchild of two local papermaking brothers, Joseph-Michel and Jacque-Etiénne Montgolfier. Their balloon consisted of a giant upside-down paper bag that rose when filled with hot air—or more correctly, hot smoke. Perhaps because the brothers actually wanted to see the gas that lifted their invention, they made sure the fire beneath the bag's opening generated visible vapors. To this end, they added old shoes and spoiled meat to the straw-fed fire that lifted the balloon.

The stench from the Montgolfier's fire became such a feature of subsequent launches that, at a royal exhibition, Louis XVI and Marie-Antoinette had to step back while the balloons inflated, lest their olfactory sensibilities be violated.

As I understand it, you didn't see the first Montgolfier hot air balloon demonstration because it occurred hundreds of miles from Paris. Lavoisier, however, did witness the event and described it in writing. Two months later, on August 27, 1783, you watched the French physicist Jacques Alexandre César Charles send aloft in Paris the first balloon filled with Cavendish's hydrogen, which he generated by pouring oil of vitriol (*sulfuric acid*) over iron filings.

Sir Joseph Banks, President of the Royal Society, wrote to you before the Charles balloon lift-off: "We are anxious here to know the event of Mr. Montgolfier's experiment and that of his competitor should you be as much inclined to philosophical amusements as we wish you to be you may possibly find time to give me a line concerning them."

Three days after the launch, you described the event to Banks: "A hollow globe 12 feet diameter was formed of what is called in England oiled silk, here *Taffetas gomme*, the silk being impregnated with a solution of gum elastic in linseed oil, as is said. The parts were sewed together while wet with the gum, and some of it was afterwards passed over the seams, to render it as tight as possible. It was afterwards filled with the inflammable air that is produced by pouring oil of vitriol upon filings of iron, when it was found to have a tendency upwards so strong as to be capable of lifting a weight of 39 pounds, exclusive of its own weight which was £25, and the weight of the air contained."

An early Montgolfier balloon carrying two passengers

You reported, "it is supposed that not less than 50,000 people were assembled to see the experiment." You went on to describe how the balloon was cut loose from it's mooring and drifted off into the clouds, landing four leagues away. After mentioning how the balloon ascent amused the multitudes, Dr. Franklin, you wrote these prophet words: "But possibly it may pave the way to some discoveries in natural philosophy of which at present we have no conception."

Not to be outdone, the Montgolfiers launched in Paris on September 19, 1783. The brothers placed a sheep, a cock and a duck on board to see if living creatures could survive the ordeal. When the balloon landed miles away, the country folks "attacked it with stones and knives," according to your report.

On November 21, you and others watched from your Passy terrace as the Montgolfiers released a huge balloon with a capacity of 50,000 cubic feet of hot air and a 1500-pound lift. On board was Jean Francis Pilatre de Rosier, the first human to fly in an aircraft. (He later died ballooning.) You wrote Banks, "This experiment is by no means a trifling one. It may be attended with important consequences that no one can foresee."

You also co-authored the official report of the flight. While observing the launch, legend says you were asked, "What good is this?" You replied, "What use is a newborn babe?" Did you actually say that, Dr. Franklin? The matter is of interest to historians because the response is often used today about any new invention.

At least a dozen letters passed between you and others about balloon experiments. Your Viennese friend Jan Ingenhousz saw interesting potential for balloons: "I can not look upon those balloons but as one of the greatest discoveries of natural philosophy; a discovery big with the most important consequences and capable of giving a new turn to human society, of overturning the whole art of conducting wars." No ship at sea would be safe, he claimed, from overhead bombardment. He asked you, "do not you think that this discovery will put an end to all wars and thus force monarchs to perpetual peace or to fight their own quarrels among themselves in a duel?"

Your reply has proven far too optimistic, Sir, especially in light of subsequent events: "It appears as you observe, to be a discovery of great importance...convincing sovereigns of the folly of wars...since it will be impracticable for the most potent of them to guard his dominions."

In that letter you mentioned altitude controls for the hydrogen balloons: A balloonist could dump sand ballast if the balloon started to descend and release a valve at the top of the bag if the balloon started to climb too rapidly.

It didn't take long for news of ballooning to reach America. You soon received a May 24, 1784 letter from Declaration signer Francis Hopkinson stating, "We have been diverting ourselves with raising paper balloons by means of burnt straw to the great astonishment of the populace."

Anticipating modern lighter-than-air ships, Hopkinson, an outstanding scientist in his own right, proposed a way of controlling the horizontal movement of lighter-than-air ships: "let the balloon be constructed of an oblong form, something like the body of a fish, or a bird...and let there be a large and light wheel in the stern, vertically mounted—this wheel should consist of several vanes...If the navigator turns this wheel swiftly round, by means of a winch, there is no doubt but it would (in a calm at least) give the machine a progressive motion..." (We call such a device a *propeller*.)

A year before the Hopkinson communication, you thought about controlling a balloon's horizontal movement: "These machines must always be subject to be driven by the winds. Perhaps mechanic art may find easy means to give them progressive motion in calm; and to slant them little in the wind."

In January 1785, Jean Pierre Blanchard and Dr. John Jeffries (an American) made history by crossing the English Channel from Dover to Calais in a hydrogen-filled balloon. They measured temperature during their ascent, learning that air gets colder at a fixed rate as altitude increases. On board they carried the first *airmail letter*, a brief note from your son William to his son Temple.

In 1932, a Swiss *aeronaut* named Auguste Piccard ascended to 10 miles in a pressurized cabin, carried aloft by a hydrogen balloon. Sixty-seven years later, his grandson and an American became the first balloonists to circumnavigate the globe by riding the winds; the journey took 19 days, 21 hours, and 55 minutes.

As I mentioned, both you and Ingenhousz foresaw the military potential for balloons, a concept that occurred to war planners as well. While aerial bombardment from balloons proved impractical—because a single bullet could bring down an airship—observation balloons played a role in warfare from 1794 until 1962.

A German Count named Ferdinand von Zeppelin went aloft while visiting an American military balloonist. He soon began work on steerable airships, with propellers powered by the newly developed internal combustion engines and rudders in the horizontal and vertical planes. His *Zeppelins* had a lightweight metal frame, hydrogen-filled balloons for lift, and a passenger compartment secured to the frame.

Transoceanic travel on huge Zeppelins became fashionable in the 1920s and 1930s with regular runs between Europe and North or South America, a trip that took 8-10 days. The great era of airship travel ended suddenly in 1937 when a hydrogen-filled Zeppelin exploded while landing.

The inconvenience of feeding a fire to sustain lift caused hot-air balloons to lose popularity shortly after the Montgolfiers died. In October 1960, however, the first "modern" hot-air balloon was developed; it uses a flammable gas to feed the fire. Now, colorful hot-air balloons once again dot skies around the world.

Passenger balloons, although pleasant enough for a day's outing, move forward at the speed of a sailing vessel and are easily buffeted by the wind. A great invention, however, increased the speed of air travel a hundred-fold over that possible with balloons.

In the early 1900s, two American brothers studied the flight of birds and kites and everything else that moves in the air. They realized that the lifting power of a bird's wing is a natural application of the Italian physicist Daniel Bernoulli's principle: air rushes more rapidly over the curved top surface than along the flat bottom surface, resulting in diminished pressure on the top compared to the bottom, providing lift. They combined this observation with the use of a windmill-like device that propels air over the wing and invented the *airplane*. Modern versions of their contraption can carry more than 400 passengers at nearly 600 miles per hour over great distances, re-ducing the trip from America to Europe in about six hours.

About 60 years ago, scientists started to build rockets almost identical to the ones used in you era to launch fireworks, although much larger. Their intent was at first military but now such contrivances carry humans into outer space—and to the moon. Somehow, I don't think you'll be surprised to learn what the space rockets use for fuel: They burn Cavendish's hydro-gen in Priestley's oxygen to lift the vehicle off the Earth.

From outer space, our planet looks like a blue and white balloon sus-pended in a dark void. It took someone like you, Dr. Franklin, to see beyond the confines of earth-bound observation and realize that we are all riding on a balloon-like planet, depending upon other objects floating in our sky for light by day and by night.

In 1783 you wrote of the divine humor in it all: "Beings of a rank and nature far superior to ours have not disdained to amuse themselves with making and launching balloons, otherwise we should never have enjoyed the light of those glorious objects that rule our day and night, nor have had the pleasure of riding round the sun ourselves upon the balloon we now inhabit."

Jacque-Etiénne Montgolfier

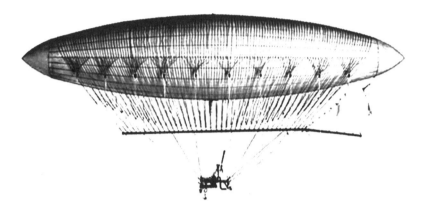

1852 Airship built by Henri Giffard

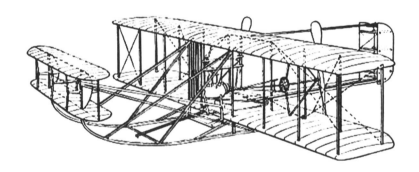

Early airplane invented by the Wright brothers of Toledo Ohio

From: "Stuart Green" <stuartgreenmd@yahoo.com>
To: "Benjamin Franklin" <dr_benjamin_franklin@yahoo.com>
Subject: **Thermometers are often badly made**

Dear Doctor Franklin:

Just as the 17th century's newly devised telescopes and microscopes exposed unknown worlds to surveillance and study, the thermometer, barometer and hydrometer revealed peculiar atmospheric features that required explanation.

Some of the most famous men in science and technology—Galileo, Ferdinand II (Duke of Tuscany), Robert Hooke, Robert Boyle, and Christian Huygens—made significant contributions to the development of the **thermometer.**

Galileo and Ferdinand II introduced the water-filled column for measuring temperature variations, while Boyle, Hooke and Huygens established the standardized points used by all thermometers—the freezing and boiling temperatures of water.

When you were eight, Gabriel Fahrenheit perfected the mercury thermometer, setting the freezing of water at 32° and human body temperature at 96°. The boiling point of water landed at 212° on his scale, although that wasn't one of his original set points. The accuracy of his thermometers made them popular in Holland and England. Other scales appeared as well, including those of René Réaumur and Anders Celsius.

By your time, thermometers were readily available; you bought one in Europe for Debbie for 18 shillings. However, you complained about their lack of reliability: "Thermometers are often badly made; I had three that differed widely from each other, tho' hung in the same place." Likewise, the proliferation of thermometer scales created confusion by the mid-18th century.

Ever practical, Dr. Franklin, you wrote *On Thermometers* to explain how to convert Fahrenheit's scale to that of Réaumur. Describing the conversion strategy we still use, you wrote, "Suppose the degree mentioned is 25 of Réaumur...Double the 25, which will give you 50 of Fahrenheit's, and to them add 32...and you will have 82, the degree of Fahrenheit's equal to 25 of Réaumur's."

The **barometer** was the second atmospheric instrument developed in the century before yours. As with the thermometer, great Galileo had a hand in the project, although his assistant Torricelli made the major contribution.

As people began recording the height of mercury in barometers, they realized that the fluctuations *preceded* weather changes, either for better (rising column) or worse (falling column). Nobody understood why. You, for instance, in a 1774 letter to the French mathematician Condorcet wrote, "The height of the barometer, by many years observation, is said to vary, between 28.59 and 30.78. The conjectures from those changes are still uncertain."

We now know that air is not a uniformly thick layer surrounding the planet, as scientists in your era believed. Instead, hills and valleys exist in the atmosphere; it would appear lumpy if visible from outer space. The atmosphere's weight beneath a high bump is much greater than that under a valley, accounting for the highs and lows on our barometers.

The lowest of the lows occurs during a hurricane, as you know. Benjamin Vaughan—the plantation owner with whom you corresponded so lucidly about lead poisoning—wrote to you in 1788 about a hurricane in Haiti, comparing it to a far worse storm that earlier hit Jamaica: "I am in that part of Hispaniola that suffered most from the hurricane last August. It wasn't however to compare with a Jamaica one, taking the barometer as the gauge and certainly there cannot be a more accurate one. This sunk but 4 Lines here, whereas it sunk more than 2 Inches in Jamaica in those disastrous misfortunes."

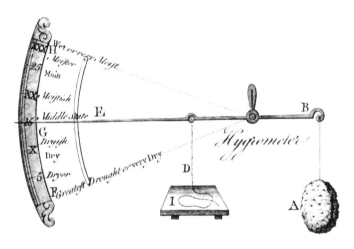

18th-century wet wool hygrometer

The third instrument essential to atmospheric science, the **hygrometer**, which measures air's moisture, was devised in the 17th century. The first hygrometer appeared a hundred years before either the thermometer or the barometer. A German amateur scientist, Cardinal Nicholas de Cusa, explained how to assess moisture in the air by using a bundle of wool: when weather turns damp, wool becomes heavier; when weather dries, wool lightens. This concept of measuring atmospheric moisture with organic material proved durable indeed. Robert Hooke, for example, attached "beard of wild oak" to a dial and started recording the changes he observed.

René Réaumur

In 1751, you grumbled about the low reliability of moisture measuring instruments: "As to hygrometers, there is no good one yet invented. The cord is as good as any, but like the rest it grows continually less sensible by time, so that the observations of one year cannot be compared with those of another by the same Instrument."

The Swiss scientist Jean du Luc designed the best hygrometer of your time by incorporating whalebone as the reactive substance. At least half a dozen correspondents told you about the device, so you might remember it—if you've memory left for such details.

For instance, recall this correspondence from British instrument maker Edward Nairne: "Mr. Du Luc shewed me an hygrometer it was made of a thin piece of whale bone about nine inches long was kept strait by a line fastened to it going over a pulley the other end of which line was fastened to a spiral spring. He observed that it was the only substance he had ever met with that would always return to the same length when soaked in water. It altered in its length about one inch from extreme moisture to extreme dryness." Five years later, Nairne sent you two of du Luc's devices for the American Philosophical Society's scientific instrument collection.

In the 19th century, a more accurate way of measuring humidity appeared, based on the relationship between two adjacent thermometers, one with an ordinary dry bulb and the other with a bulb surrounded by a wet gauze.

The wet bulb effect occurs because evaporation of a liquid cools its surface, a fact discovered during your era by one of your acquaintances, Dr. William Cullen. (You met him during your trips to Scotland.) You described his discovery thus: "Dr. Cullen, of Edinburgh, has given some experiments of cooling by evaporation…by repeatedly wetting the ball of a thermometer with spirit [ether], and quickening the evaporation by the blast of a bellows…ice gathered in small spicula round the ball, to the thickness of near a quarter of an inch."

Explaining the phenomena in 18th-century terms, you assumed that the mercury did "lose the fire it before contained…took the opportunity of escaping, in company with the evaporating particles of the spirit, by adhering to those particles." (Fire, during your time, was an imponderable substance and is best thought of as *heat* in my terms.)

You quickly realized that cooling by evaporation explained many common practices: "Even our common sailors seem to have had some notion of this property," you informed a correspondent. While at sea, you "observed one of the sailors, during a calm in the night, often wetting his finger in his mouth, and then holding it up in the air, to discover, as he said, if the air had any motion, and from which side it came; and this he expected to do, by finding one side of his finger grow suddenly cold, and from that side he should look for the next wind."

You laughed at the sailor's maneuver "as a fancy." But after seeing the effect of evaporation during Cullen's thermometer experiment, you asked, "May not several phenomena, hitherto unconsidered, or unaccounted for, be explained by this property?"

The benefit of sweating was a perfect example: "During the hot Sunday at Philadelphia, in June 1750, when the thermometer was up at 100 in the shade, I sat in my chamber without exercise, only reading or writing, with no other cloaths on than a shirt," you explained to John Lining. Recognizing the value of perspiration, you wrote, "in this situation, one might have expected, that the natural heat of the body 96, added to the heat of the air 100, should jointly have created or produced a much greater degree of heat in the body; but the fact was, that my body never grew so hot as the air that surrounded it."

You repeated the story to several correspondents, adding to one, "So it should seem, that a naked man standing in the wind and repeatedly wet with strong spirit, might be frozen to death on a summers day." You couldn't explain how evaporation cools: "I think none of our common philosophical principles will serve us in accounting for this."

Obviously impressed with folk knowledge of evaporation's cooling effect, Dr. Franklin, you wrote to another friend, "it has been long known in the East…the custom of travelers carrying their water in flasks covered with wet woolen cloth, and hung to the pommels of their saddles, so as that the wind might act upon them in order to cool the water."

From: "Stuart Green" <stuartgreenmd@yahoo.com >
To: "Benjamin Franklin" <dr_benjamin_franklin@yahoo.com>
Subject: **The dissolved water in the air**

Dear Doctor Franklin:

Measuring **rainfall**, another important weather statistic, has engaged human interest for millennia. It's the easiest of all meteorological data to acquire. The variations in raindrop size, however, puzzled thinkers for centuries. Aristotle believed that the largest raindrops formed closest to the ground, a notion that persisted up to your time. Unfortunately, an ill-conceived experiment perpetuated the myth.

Dr. William Heberden, your co-author on the smallpox pamphlet, in 1766 set up three rain gauges in Westminster, each at a different height. The lowest was at ground level, the middle one at chimney height, and the highest on the roof of Westminster Abbey. He observed that the amount of rainwater collected in the gauges increased the lower one got: 12 inches on the abbey's roof, 18 inches at the chimney and 22 inches at ground level.

We now know that swirling wind eddies caused the rain to blow sideways, missing to upper collectors at the top of the Abbey and chimney, accounting for the difference. Heberden, however, confident of his results, suggested that an electrostatic charge on raindrops attracted additional moisture as they fell.

You offered Dr. Thomas Percival two possible explanations for Heberden's observation. You based one proposal on the way water droplets collect on the outside of a cool vessel on a muggy day. Perhaps, you speculated, a cold raindrop attracts to its surface "the dissolved water in the air." A second explanation, as Heberden suggested, might relate to electrostatic charges, just as dust collects from the air on "an electrified body left in a room for some time." However, you didn't think you had enough information to offer a good hypothesis.

Although you considered "Dr. Heberden to be very accurate," you said, "I think we want more and a greater variety of experiments in different circumstances, to enable us to form a thoroughly satisfactory hypothesis." In particular, you wanted to be sure that a *mathematical* relationship existed between the heights of the rain gauges and the amount of water each contained before committing to an explanation of Heberden's findings. As you put it, "whether in the same place where the lower vessel receives nearly twice the quantity of water that is received by the upper, a third vessel placed at half the height will receive a quantity proportionable."

Indeed, Dr. Franklin, a greater variety of experiments in different locations were eventually made, correcting Heberden's erroneous data.

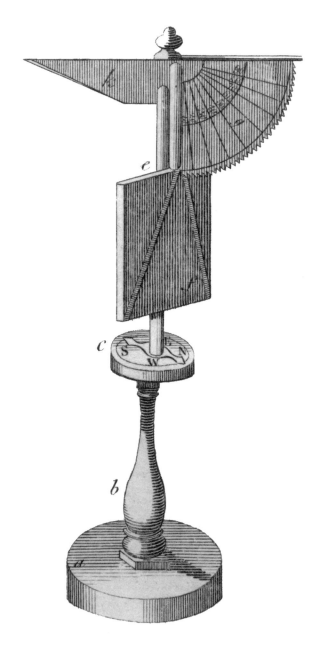

18th-century wind gauge

From: "Stuart Green" <stuartgreenmd@yahoo.com>
To: "Benjamin Franklin" <dr_benjamin_franklin@yahoo.com>
Subject: **As easy as pissing abed**

Dear Doctor Franklin:

Most people are unaware of your contributions to atmospheric science. You confirmed, of course, the long held suspicion that lightning was a form of electric discharge. That discovery alone assured you a place in weatherman firmament.

Your *Poor Richard's Almanack* carried weather forecasts based on old wives' tales and common climate knowledge. It appears that you never resorted to astrology for your predictions and even mocked yourself for the guesses: "Ignorant men wonder how we astrologers foretell the weather so exactly, unless we deal with the old black Devil. Alas! 'tis as easy as pissing abed."

In truth, Sir, you made a distinctly scientific contribution to weather prediction in a most ingenious way. Like everyone of your time, you assumed that storms came to Philadelphia from the northeast, because that's the way the wind blew in. Sometime in the 1740s, however, you discovered that storms actually travel from southwest to northeast.

On two separate occasions you wrote to friends explaining how you figured out this apparent contradiction: "What first gave me this idea, was the following circumstance. About twenty years ago...we were to have an eclipse of the moon at Philadelphia, on a Friday evening, about nine o'clock. I intended to observe it, but was prevented by a north-east storm, which came on about seven, with thick clouds as usual, that quite obscured the whole hemisphere."

You later read in a Boston newspaper about the same storm, yet were surprised that "the eclipse had been well observed there." "This puzzled me," you wrote, because you assumed a storm would start <u>sooner</u> in Boston than in Philadelphia. So you wrote to brother James in Boston who informed you, "the storm did not begin with them till near eleven o'clock, so that they had a good observation of the eclipse." You took the next step and requested information from correspondents in other colonies about the storm's relationship to the eclipse. You found "the beginning to be always later the farther north-eastward...about an hour to every hundred miles." (The actual speed of travel for storms, Dr. Franklin, is about 10-15 miles per hour.)

By comparing the storm's onset times in different locations to a fixed celestial event—a lunar eclipse—you correctly determined the North American continent's weather pattern years before meteorological records and weather balloons confirmed your observation.

In general, you and your fellow 18th-century philosophers knew less about meteorology than any other natural science. While alchemists were methodically studying acids, alkalis, metals and calyxes, those interested in weather phenomena found themselves pondering the same questions posed by the ancients: What holds up the clouds? Where does rain come from? Why do hailstones fall in the summer?

Remarkably, you had something to say about each of these subjects.

Recall that the four-element system helped explain weather phenomena for 1500 years. The Greeks observed that a puddle of water evaporated faster on a sunny day than on a cloudy one. The sun, therefore, must be similar to the fire that "evaporates" boiling water in pot. Fire was thus associated with heavenly bodies.

In the 17th century, perceptive thinkers began calling for experiments and data to support meteorological conjectures. A particularly intriguing question was how water got up into the clouds. After all, water is a liquid and aren't liquids heavier than air?

In 1765, Dublin professor Hugh Hamilton proposed that water vapor *dissolves* in air, just as salt dissolves in water. After Hamilton presented his ideas at the Royal Society, someone remembered that ten years earlier you had submitted a manuscript containing the same hypothesis, based on the identical reasoning. The Society immediately published your article—with an apology.

Unfortunately, you started your submission by repeating the erroneous Newtonism that all particles of air are mutually repellant. (Air, when you wrote the paper, was viewed as a single substance, not a mixture of gases.)

Next, you stated the principle that "air and water mutually attract each other. Hence water will dissolve in air, as salt in water." Carrying the analogy further, you pointed out that water will continue to dissolve salt until it can take no more, at which point the salt precipitates out of solution. Thus, "water in the same manner will dissolve in air, every particle of air assuming one or more particles of water; when too much is added, it precipitates in rain." Motion enhances the mixing of water in the air, just as stirring helps dissolve sugar or salt in water.

Earth also mixes with air, according to your viewpoint. Dust swirled up by a horse's hoof rapidly diffuses into the air, making a cloud as "big as a common house." The dust cloud, you proclaimed, wasn't enlarged by the horse's motion or the wind. Instead, warm air rising from the sun-heated ground lifted the dust particles: "Quantities of dust are thus carried up in dry seasons: Showers wash it from the air and bring it down again."

Your concept of air rising above ground heated by solar radiation was quite correct. You mentioned this point in several communications to friends and correspondents, rightly concluding that wind formed as cold air moved in to replace rising warm air.

Since the upper atmosphere is cooler than the lower air, water vapor carried aloft by the ground-heated air will condense out, you concluded, just as water collects from the air on cold glass. You explained cloud formation thusly: "The cold air descending from above, as it penetrates our warm region full of watery particles, condenses them, renders them visible, forms a cloud thick and dark...." A small cloud, according to your view, "increase its bulk, it descends with the wind and its acquired weight, draws nearer the earth, grows denser with continual additions of water, and discharges heavy showers." The rising movement of warmed air and its replacement with heavy colder air account for the high velocity wind phenomena that occur during the summer. As you put it: "...the heaviest part descends first, and the rest follows impetuously. Hence gusts after heats, and hurricanes in hot climates."

Luke Howard

Before any scientific discipline can make progress, researchers in the field must speak the same language. A major advance in meteorology occurred in 1802 when Luke Howard, a British pharmacist, gave names to the different kinds of clouds. Howard was born into a Quaker family in 1772—entering the world in the same month (November) you supported John Pringle's successful campaign for presidency of the Royal Society. By selecting Latin terms—cumulus, stratus, cirrus, and nimbus—for distinct cloud formations, and noting that each type occupies discrete layers of the atmosphere, Howard created a system of nomenclature that has survived to this day. He also kept accurate weather records, which he employed to make forecasts about future meteorological events. Like you, Howard had an abiding interest in lightning. He incorrectly concluded that electrical activity in clouds caused rain. Nevertheless, meteorologists around the world recognize Luke Howard as a pioneer of their field.

A groundbreaking discovery in meteorology, made by someone also born during your lifetime, occurred after a violent hurricane struck New England in 1821. A young apprentice saddle-maker named William C. Redfield noticed that the trees blown down near his home in Middletown, Connecticut faced northeast, whereas 70 miles away in Massachusetts, the fallen foliage pointed southwest. Redfield kept this observation to himself for ten years, until he happened by chance to meet a university professor on a steamship. In the interim, Redfield had become a successful businessman and self-taught amateur scientist.

The professor encouraged Redfield to publish his theory that hurricanes are gigantic counterclockwise rotating whirlwinds, more than 100 miles across. Redfield did just that, and included a speculation—since proven correct—that hurricanes below the Equator spin in the opposite direction.

Hurricane over Bermuda, viewed from above
Image from NASA

Redfield went on make additional fruitful contributions to the analysis of hurricanes and other tropical cyclones. Perhaps even more significantly, he became founding president of the American Association for the Advancement of Science, a premier research society that still flourishes today.

In spite of gradually increasing knowledge about the behavior of winds, the nature of high and low pressure systems, and other atmospheric features, weather prediction remained more art than science. The advent of Morse's telegraph system allowed simultaneous measurements of weather conditions over continental distances to be collected in one place for analysis, but in the end, guesswork prevailed.

In 1904, a young Norwegian physicist named Vilhelm Bjerknes published a thoughtful article suggesting that, with accurate information about weather data at any point in time, and adequate understanding of the law of physics governing atmospheric change, one could successfully forecast the weather. He eventually established an institute in Bergen, Norway to develop the theoretical background necessary to achieve his goal.

Soon, students interested in meteorology flocked to Bergen for advanced training. In time, these students (including his son Jacob) dispersed to create departments of meteorology at leading universities around

Vilhelm Bjerknes

the world. And from these departments, Dr. Franklin, have emerged today's leaders in atmospheric science and weather forecasting.

From: "Stuart Green" <stuartgreenmd@yahoo.com >
To: "Benjamin Franklin" <dr_benjamin_franklin@yahoo.com>
Subject: **Whirlwinds at land, waterspouts at sea**

Dear Doctor Franklin:

Although the *Coriolis Effect* describing, among other things, the influence of the Earth's rotational inertia on the atmosphere wasn't proposed until 1835 you, Sir, offered your own remarkable version in a 1751 submission to the Royal Society.

Combining the effect of heat and motion, you stated, "The air under the equator, and between the tropics, being constantly heated and rarified by the sun, rises. Its place is supplied by air from northern and southern latitudes, which coming from parts where the earth and air had less motion, and not suddenly acquiring the quicker motion of the equatorial earth, appears an east wind blowing westward." Additionally, you claimed, "The air rarified between the tropics, and rising, must flow in the higher region North and South. Before it rose, it had acquired the greatest motion the earth's rotation could give it."

Noting how water swirls around a drain, you reminded readers that "heavy fluids descending, frequently form eddies, or whirlpools, as is seen in a funnel, where the water acquires a circular motion." Once this happens, air has to rise in a swirl to replace it: "Thus, these eddies may be whirlwinds at land, waterspouts at sea." You were therefore drawn to whirlwinds and waterspouts as confirmation of your atmospheric conjectures.

Tornado chasing today—an avocation for some, a profession for others—is the often exhilarating and occasionally dangerous pursuit of a particularly destructive atmospheric event. And you, Dr. Franklin, pioneered the practice.

In 1755, you told Peter Collinson how you chased a whirlwind across the Maryland countryside. You were on horseback, riding with your friend Col. Benjamin Tasker, son William and others, when you spotted in the valley below "a small whirlwind beginning in the road, and showing itself by the dust it raised and contained." The whirlwind was "enlarging as it came forward. When it passed by us, its smaller part near the ground appeared not bigger than a common barrel, but widening upwards, it seemed, at 40 or 50 feet high, to be 20 or 30 feet in diameter."

Your companions watched the whirlwind's progress, you told Collinson, "but my curiosity being stronger, I followed it, riding close by its side, and observed its licking up, in its progress, all the dust that was under its smaller part."

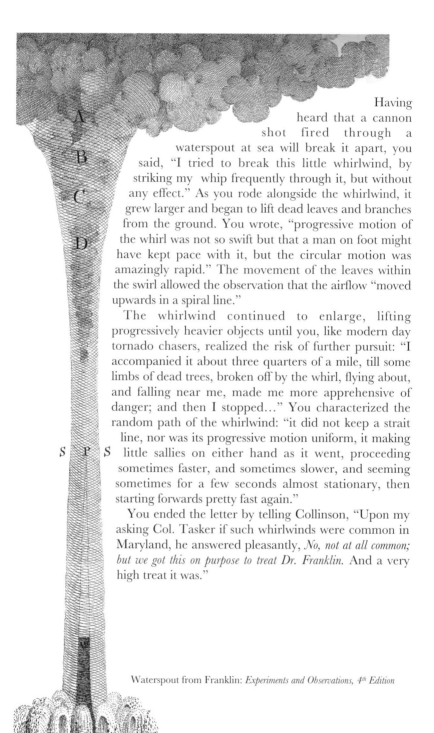

Having heard that a cannon shot fired through a waterspout at sea will break it apart, you said, "I tried to break this little whirlwind, by striking my whip frequently through it, but without any effect." As you rode alongside the whirlwind, it grew larger and began to lift dead leaves and branches from the ground. You wrote, "progressive motion of the whirl was not so swift but that a man on foot might have kept pace with it, but the circular motion was amazingly rapid." The movement of the leaves within the swirl allowed the observation that the airflow "moved upwards in a spiral line."

The whirlwind continued to enlarge, lifting progressively heavier objects until you, like modern day tornado chasers, realized the risk of further pursuit: "I accompanied it about three quarters of a mile, till some limbs of dead trees, broken off by the whirl, flying about, and falling near me, made me more apprehensive of danger; and then I stopped..." You characterized the random path of the whirlwind: "it did not keep a strait line, nor was its progressive motion uniform, it making little sallies on either hand as it went, proceeding sometimes faster, and sometimes slower, and seeming sometimes for a few seconds almost stationary, then starting forwards pretty fast again."

You ended the letter by telling Collinson, "Upon my asking Col. Tasker if such whirlwinds were common in Maryland, he answered pleasantly, *No, not at all common; but we got this on purpose to treat Dr. Franklin. And a very high treat it was.*"

Waterspout from Franklin: *Experiments and Observations, 4th Edition*

From: "Stuart Green" <stuartgreenmd@yahoo.com>
To: "Benjamin Franklin" <dr_benjamin_franklin@yahoo.com>
Subject: **Some curious and useful discoveries in meteorology**

Dear Doctor Franklin:

Two centuries before scientists proposed asteroid collisions as the cause of *mass extinctions* of ancient creatures, you realized that extraterrestrial and geological events could adversely affect climate.

Do you recall receiving honorary membership in the Philosophical Society of Manchester for your *Meteorologic Imaginations and Conjectures*? In that work, you ascribed the severe winter of 1783-1784 to a peculiar feature of that year's summer, when "there existed a constant fog over all Europe." The fog "was dry, and the rays of the sun seemed to have little effect towards dissipating it, as they easily do a moist fog arising from water." Because the fog blocked the sun's rays from reaching the ground, "heating the Earth was exceedingly diminished," and therefore "the surface was early frozen… the first snows remained on it unmelted…the air was more chilled, and the winds more severely cold."

"The cause of this universal fog is not yet ascertained," you wrote, offe r-ing two promising explanations, both equally ingenious. One possibility was smoke from burning meteors, "which we happen to meet with in our rapid course round the sun, and which are sometimes seen to kindle and be destroyed in passing our atmosphere, and whose smoke might be attracted and retained by our Earth."

Your other conjecture involved volcanic fumes: "the vast quantity of smoke, long continuing to issue during the summer from Helca in Iceland, and that other volcano which arose out of the sea near that island; which smoke might be spread by various winds over the northern part of the world."

Extinct volcanoes, from Humphreys' *Nature Display'd*

You thought it a worthy subject of inquiry "whether other hard winters recorded in history, were preceded by similar permanent and widely-extended summer fogs." If so, people could, upon experiencing such a summer fog, "take such measures as are possible and practicable to secure themselves and effects from the mischiefs" associated with unusually harsh winters.

Your hypothesis about the influence of volcanic eruptions on worldwide weather patterns has proven correct. Your Manchester manuscript, Sir, in light of present knowledge, contained noteworthy speculations into atmospheric phenomena. Based on the presence of hailstones in the summer, you concluded that "There seems to be a region high in the air over all countries, where it is always winter."

And indeed there is such a region.

Nowadays scientists divide our atmosphere into five layers. The closest to Earth, the *troposphere*, measures several miles thick and contains the weather-making features. At the top of the troposphere, the *tropopause*, your prediction was correct because it's always winter up there, with average temperatures around 60 degrees below zero on Fahrenheit's scale. A 20 mile thick *stratosphere* surrounds the troposphere, with temperatures at −80°F. The stratosphere contains the *jet stream*, which has profound influence on weather movement.

The next layer out, the *ozonosphere*, is warmer (up to +30°F) because it absorbs the sun's most harmful rays, shielding the earth.

The atmosphere cools considerably into the *mesosphere*, about 50 miles thick, down to −135°F, but the situation heats up again in the outer atmosphere, the *thermosphere*. This layer extends out to about 375 miles and, with direct sunlight, gets hot as Hell: +3000°F.

By assuming a cold layer in the atmosphere, Dr. Franklin, you correctly accounted for several weather effects. For example, modern meteorologists believe that most rain starts as snow. In 1784 you put it this way: "It is possible that in summer, much of what is rain when it arrives at the surface of the earth, might have been snow when it began its descent; but being thawed in passing thro' the warm air near that surface, it is changed from snow into rain."

As for hailstones, they start as particles of water vapor. "As soon as they are condensed by the cold of the upper regions so as to form a drop," you surmised, "that drop begins to fall...If it freezes into a grain of ice, that ice descends. In descending both the drop of water and the grain of ice, are augmented by particles of the vapour they pass thro' in falling, and which they condense by their coldness and attach to themselves."

You also accounted for the large size of hailstones by assuming a very cold initial temperature, sufficient to "freeze all the Mass of Vapour condensed round it, and form a Lump of perhaps 6 or 8 ounces in weight." We now realize that the biggest hailstones make several up and down trips through thunderclouds. They start out rather small and, upon falling, are lifted back up by rising warm air currents. Each vertical passage enlarges the hailstones with frozen water vapor until the uprising air can no longer support them, so they fall.

Slicing a hailstone in half reveals a layering of ice—the number of layers equals the number of passages through the clouds. Recent research has revealed that hailstones are formed in a number of different ways, including the mechanism you proposed in 1784.

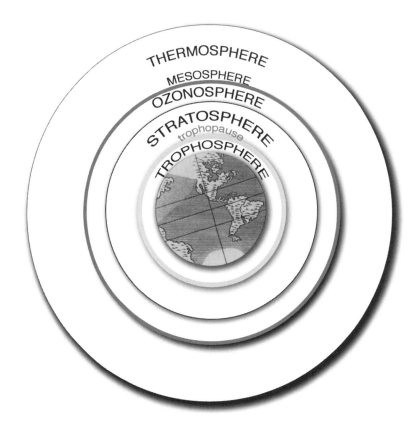

Layers of Earth's atmosphere (not to scale)

By the time you watched the balloon launches in Paris, fairly reliable atmospheric instruments were available. Moreover, balloonists took the devices aloft on the earliest ascents. In describing a hydrogen balloon that Charles planned to launch later in 1783, you wrote of it: "A very handsome triumphal car will be suspended to it…There is room in this car for a little table…in which they can write and keep their Journal, that is, take notes of every thing they observe, the state of their thermometer, barometer, hygrometer, &c., which they will have more leisure to do than the others, having no fire to take care of."

Aside from the occasional meteorological data generated by the infrequent balloon launches, you also recognized the value of accurate, long-term terrestrial weather records.

By the mid 19th century, every technologically advanced nation had an organized system for gathering meteorological data. Weather forecasting became a more accurate science following the development of more sophisticated measuring instruments and especially, the 1844 invention of the telegraph

Today, thanks to electronic transmittal of information, weather prediction is a worldwide enterprise, with accuracy improving all the time. We now realize that human activity adversely affects climate. Our planet is gradually warming as a result of burning natural fuels, which changes the composition of the atmosphere. This problem, unless reversed, may have dire consequences.

In a letter to Dr. Thomas Bond, you mentioned the life-long weather journals Scottish surgeon Alexander Small donated to the American Philosophical Society, with you as the intermediary for the transfer. You told Bond of Small's "scheme for noting the variations of the barometer, and comparing them in different and distant places, which he recommends to be used by the members of the Society that inhabit different provinces, as he conceives that some curious and useful discoveries in meteorology may thence arise."

And indeed they have.

Thomas Bond

From: "Stuart Green" <stuartgreenmd@yahoo.com>
To: "Benjamin Franklin" <dr_benjamin_franklin@yahoo.com>
Subject: **Different degrees of heat**

Dear Doctor Franklin:

Your interest in heat and color, which I touched on earlier, has stimulated dispute among historians, with some asserting that *you* discovered the relationship between heat absorption and color.

Polly Stevenson Hewson

Do you recall writing to Polly Stevenson in 1761 about dark and light clothes? To refresh your memory, here's what you said: "Let me mention an experiment you may easily make yourself. Walk but a quarter of an hour in your garden when the sun shines, with a part of your dress white, and a part black; then apply your hand to them alternately, and you will find a very great difference in their warmth. The black will be quite hot to the touch, the white still cool."

You next suggested that she focus a burning glass on blank white paper and on paper with black printing on it. "If it is white, you will not easily burn it," you told her, "but if you bring the focus to a black spot or upon letters written or printed, the paper will immediately be on fire under the letters."

You also mentioned that fabric dyers claim that black cloth dries faster in the sun than white cloth, and that the heat of a fire "sooner penetrates black stockings than white ones." Lastly and most importantly, you informed Polly, "beer much sooner warms in a black mug set before the fire, than in a white one, or in a bright silver tankard."

I'm particularly intrigued by what you called "my experiment," namely, putting colored fabric patches onto fresh snow. You told Polly, "black being warmed most by the sun was sunk so low as to be below the stroke of the sun's rays; the dark blue almost as low, the lighter blue not quite so much as the dark, the other colours less as they were lighter; and the quite white remained on the surface of the snow, not having entered it at all."

"May we not learn from hence," you proclaimed, "that black cloaths are not so fit to wear in a hot sunny climate or Season as white ones...That summer hats for men or women, should be white, as repelling that heat which gives the headaches to many, and to some the fatal stroke that the French call the *Coup de Soleil?*" You offered other suggestions as well.

I've read that you conceived the wintertime colored fabric experiment as early as 1729 in conjunction with Joseph Breintnal, a fellow Junto member. Breintnal, in his 1936 notes on the subject, credits you with the idea that dissimilar colors absorb different amount of the sun's heat. The eminent Harvard historian of science I. B. Cohen, however, has pointed out that several prior researchers, including Newton, Boerhaave, and Boyle, earlier commented about the phenomenon.

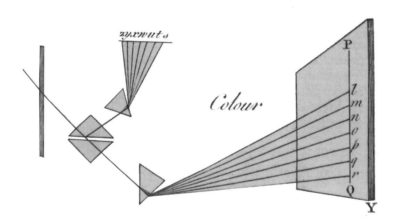

Prisms splitting and refracting a light beam, from Middleton's *Dictionary*

Today, we understand that white, a combination of all colors, *reflects* sunlight, whereas other colors absorb what they lack from the solar (or a candle's) spectrum, causing warmth.

I've got more to say about radiant heat, a subject of much interest to you, in my next email, but I must end now and go for my morning walk, having decided to follow your advice about exercising.

From: "Stuart Green" <stuartgreenmd@yahoo.com>
To: "Benjamin Franklin" <dr_benjamin_franklin@yahoo.com>
Subject: **The author's description of his Pennsylvania fire-place**

Dear Doctor Franklin:

To continue about heat, here's an update on your Pennsylvanian Fire-place.

Realizing that fireplaces set into a wall heated homes inefficiently, you invented a rectangular metal unit that extended into a room. By writing *An Account of the New-invented Pennsylvanian Fire-Places* you made a compelling argument for your protruding device.

You were particularly impressed with the earlier ideas of French inventor Nicholas Gauger. In 1713 he devised a metal firebox set away from the wall. His invention contained hollow chambers in contact with the fire but sealed from the fumes. Air entered and exited those chambers through vents, heating the room without any smoke, which went up the chimney.

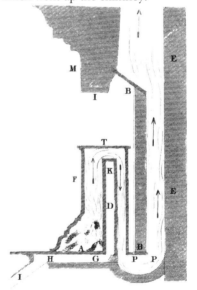

I regret to remind you that your original design never sold well. Your brother Peter tried retailing them in Rhode Island; he inventoried 12 fireplaces but sold only two. Problems with your fireplace followed installation. Masons mounted them with the metal plate that formed the back wall of your fireplace becoming the lower chimney's front wall.

Smoke in your system first had to travel *downwards* through a heating baffle before it could rise into the chimney. Since the existing stone chimney was exterior to the house, its cold air was *heavier* than the fireplace's and would resist displacement unless your fireplace was stoked quite hot. As your fireplace cooled down, smoke took the path of least resistance and entered the room.

Although you may not remember it now, you knew of these difficulties, having mentioned them in letters. You believed that many installations involved modification of your original design, causing the smokiness.

In 1787, while describing a newer *charcoal* fireplace, you wrote, "The author's description of his Pennsylvania fire-place, first published in 1744, having fallen into the hands of workmen in Europe, who didn't, it seems, well comprehend the principles of the machine, it was much disfigured in their imitations of it..." Your attempt to improve upon your original design —adding a stone vase to radiate heat as it sat above a firebox—never caught on.

While you were still alive, many fireplace modifications—forgive me, Sir, *improvements*—appeared in Philadelphia and elsewhere. Your friend David Rittenhouse enhanced your fireplace by placing the chimney directly *above* the firebox, thereby eliminating the baffle and downward flowing smoke. His device vented fumes outside via a pipe. Rittenhouse's design became quite popular; some remain in use.

What's remarkable about your legacy, Dr. Franklin, is that Americans are so convinced of your genius that *any* heater or fireplace set away from a wall is now called a *Franklin stove*, even though it bears no resemblance to your invention. Nevertheless, by your persistent interest in fuel economy, you've assumed the mantle of inventor for all efficient fireplaces and central heating units.

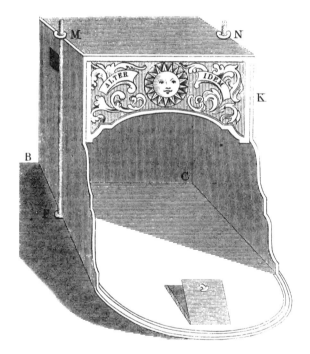

The Pennsylvania Fireplace

From: "Stuart Green" <stuartgreenmd@yahoo.com>
To: "Benjamin Franklin" <dr_benjamin_franklin@yahoo.com>
Subject: **I soon became a thorough Deist**

Dear Doctor Franklin:

I sometimes wonder how you managed to come up with so many valid insights into the workings of nature. When you wrote about lightning, geology, ocean salinity, weather prediction or dozens of other topics, your clear-eyed sagacity shines through. I've concluded that rejecting your father's faith opened your mind to rationality, although I'm not sure why you turned your back on Calvinism's fundamental tenets.

Clearly, you questioned several of Calvinism's basic canons. Isn't true that Calvinists believe that God "elected" those souls to be saved at the beginning of time? That Christ's self-sacrifice cancels the sins of the elect, offering them grace? That everyone else was damned, born in sin, corrupt or morally depraved? Did you tire of those weekly Calvinist sermons, which emphasized sin, damnation, and punishment?

Were you mad at your father for some real or imagined slight? Perhaps you resented being indentured as a printer's apprentice to your brother James for so many years.

Rest assured, Sir, you're not the only person who disliked going to church. You wrote in your *Autobiography* that you began "evading as much as I could the common attendance on publick worship, which my father used to exact of me when I was under his care." You also recorded what happened next: "I was scarce 15 when…I began to doubt of revelation itself. Some books against Deism fell into my hands…they wrought an effect on me quite contrary to what was intended…In short, I soon became a thorough Deist."

As a Deist, you rejected Scripture and chose instead to understand the natural world through rational analysis of known facts.

Deism, I understand, denies all divine revelation, holy books, or miracles. It holds that Jesus, Buddha, Mohammed were mortals. Most Deists also doubt special providence, meaning that God does not become involved with human day-to-day affairs. Likewise, the clergy serves no purpose. Nevertheless, most Deists believe in the soul's immortality.

As you got older, you gradually drifted back towards your ancestors' creed, as do many others when they mature. Some religious thinking became more acceptable to you: a Creator of the universe; divine providence; immortality of the soul. In spite of these concessions, however, you never looked to the Bible to explain anything, nor concerned yourself with seeming conflicts between faith and knowledge. Cut free from theological dogma, you sought and found answers in the realm of reason, much to your eternal credit—and the everlasting benefit of mankind.

We sometimes say, "In the land of the blind, the one-eyed man is king." In your times and still today, many people are blinded by their faith, willingly rejecting irrefutable evidence if it conflicts with their preset beliefs. I often quote the way you put it:

> In the history of mankind, every...sect supposing itself in posses-
> sion of all truth, and that those who differ are so far in the wrong.
> Like a man traveling in foggy weather, those at some distance before
> him on the road he sees wrapt up in the fog as well as those behind
> him, and also the people in the fields on each side; but near him all
> appears clear. In truth, he is as much in the fog as any of them.

We are all in that fog, Sir, though few admit it.

EXPERIMENTS

AND

OBSERVATIONS

ON

ELECTRICITY,

MADE AT

PHILADELPHIA in AMERICA,

BY

BENJAMIN FRANKLIN, L. L. D. and F. R. S.

To which are added,

LETTERS and PAPERS

ON

PHILOSOPHICAL SUBJECTS.

The Whole corrected, methodized, improved, and now firſt collected into one Volume,

AND

Illuſtrated with COPPER PLATES.

LONDON:

Printed for DAVID HENRY ; and fold by FRANCIS NEWBERY,
at the Corner of St. Paul's Church-Yard.

MDCCLXIX.

Fourth Edition of *Experiments and Observations*

From: "Stuart Green" <stuartgreenmd@yahoo.com >
To: "Benjamin Franklin" <dr_benjamin_franklin@yahoo.com>
Subject: **A fool and a madman**

Dear Doctor Franklin:

Now would be a good time to tell you something your email correspondent. I reside in a region called in your time Nuevo Albion, which entered the union as *California* sixty years after your 1790 death. I'm a professor in the School of Medicine in a city named Irvine, a part of my state's University system. I teach bonesetting and related skills. I'm also on the Board of Directors of *The Friends of Franklin*, an organization dedicated to disseminating knowledge about your life and times.

You might wonder why I chose to bring you up to date on technological, scientific and medical advances since your encaskment. Here's the story.

Several years ago I became curious about the origin of placebos for medical research, so I used my computer (similar to the one you're now reading) to *surf the web*; that is, look up information accessible on millions of interconnected computers around the world.

On the *website* of The Royal College of Physicians of Edinburgh (recall many of your friends belonged), I found a history of sham procedures and intentional patient ignorance in medical research, starting with your 1784 Royal Commission inquiry of Dr. Mesmer.

I read several Franklin biographies and noticed that although every author mentioned the investigation, none recognized its significance in the history of clinical science. Before reading those biographies, I knew that you flew a kite in the rain to prove the relationship of lightning to electricity, and that you invented the lightning rod and a few other clever gadgets. Likewise, I knew of your central role in our nation's formation, but not much else. I thus decided to learn more about your wide-ranging interests in the medical and scientific matters.

I soon learned what every historian—but few Americans—grasp: Your political influence evolved from your reputation as scientist. Your electrical experiments in the late 1740s, and the subsequent London publication of your *Experiments and Observations on Electricity made at Philadelphia in America* brought international renown. Fellow colonists realized that your growing reputation could prove useful in transatlantic negotiations, first with the Mother Country and later, after we broke with Britain, with France. Without your research, you'd have remained a retired printer of waning influence until your dying day. Without your stature, our nation might've remained subject to the English Crown.

During your first lifetime, four editions of *E & O* appeared, the last containing many of your non-electrical ideas and insights. Several translations of the fourth edition appeared, the French version produced by Jacque Barbeu-DuBourg. While reviewing your correspondence to him, I came across your letter about revitalizing flies drowned in Madeira. It occurred to me that you might someday try the experiment on yourself, as you did with so many other projects. And why not? You were close to death in early 1790, with an excruciatingly painful bladder stone and great respiratory distress. Under the circumstances you could, with a barrel of wine, simultaneously end your misery and preserve yourself for posterity.

Today Dr. Franklin, some people facing death due to illness choose to have their bodies frozen solid, to be thawed out in the future—after doctors discover a cure for their fatal disease. People pay handsome sums to enterprises that preserve corpses in this manner. Such companies often quote your Barbeu-DuBourg letter as an endorsement of their practices.

To avoid being called murderers, modern body-freezers must wait until a doctor certifies death before taking possession of the cadaver. Thus, you weren't alone in suspended mortality; thousands of likeminded individuals wait being "recalled to life" as you put it.

Skeptics enjoy pointing out that freezing destroys brain tissue; they say the corpses will never revitalize. However, so does alcohol, so on reflection, your experiment shouldn't work either.

Based on the foregoing premise, I labor on, assuming I'll soon bask in your reflected glory. I dare not share my assumption most friends, lest they think me an idiot or crazy. However, from time to time I've discussed my emails to you with two knowledgeable acquaintances, Stanley Finger and Marty Margold. While neither holds much hope of your imminent revitalization, both have offered many helpful suggestions about the content and style of our one-way correspondence.

You once wrote: "the difference between a fool and a madman is said to be, that a fool reasons wrong from the right premises, the madman right but from wrong premises." If you, indeed, acted upon your Madeiraed fly conjecture, than my effort on your behalf will prove fruitful. However, if I've erred about your *mort-faux*, I've nevertheless learned much while preparing these emails.

MARINE SCIENCE & OCEANOGRAPHY

⊠	The edges of the Gulph Stream	259
⊠	Divide the hold of a great ship into a number of separate chambers	261
⊠	One of Cook's vessels	267
⊠	Draw a boat in deep water	271
⊠	Their waters are fresh quite to the sea	273
⊠	The effect of oil on water	275
⊠	Propelling boats on the water, by power of steam	279
⊠	A sooty crust on the bottom of the boiler	283
⊠	A large boat rowed by the force of steam	287
⊠	The types he has been reading all his life	291

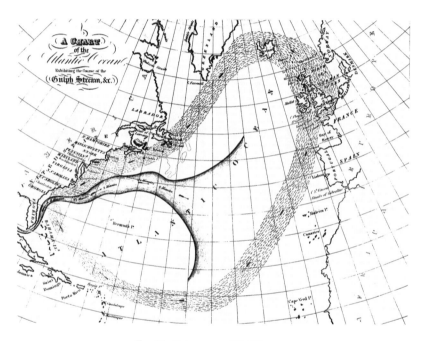

Franklin-Folger map of Gulf Stream

From: "Stuart Green" <stuartgreenmd@yahoo.com>
To: "Benjamin Franklin" <dr_benjamin_franklin@yahoo.com>
Subject: **The Edges of the Gulph Stream**

Dear Doctor Franklin:

There exists today the common misconception that you "discovered" the *Gulf Stream*. Historians, however, give that honor to the Spanish explorer Ponce de Leon. Although by the early 1700s both Spanish and American sailors knew of a warm current off America's coast, the British, you realized, seemed unaware of it.

Your role as Deputy Postmaster for the North American colonies first brought the Gulf Stream to your attention. For an unknown reason, certain trans-Atlantic mail routes took two weeks longer than others for virtually the same distance traveled.

You advised the British Postal Secretary that your cousin (Nantucket sea captain Timothy Folger) described how "whales are found generally near the edges of the Gulph Stream, a strong current so called which comes out of the Gulph of Florida, passing northeasterly along the coast of America, and then turning off most easterly..." New England whalers, you learned, are "better acquainted with the course, breadth, strength and extent of the same, than those navigators can well be who only cross it in their voyages to and from America."

Whenever American whale hunters encountered British mail packets having trouble making headway in the Gulf Stream, they told the English captains about the northeasterly flowing current. The haughty Brits, you said, didn't heed the advice because "they were too wise to be counseled by simple American fishermen."

Since you thought, "it was a pity no notice was taken of this current upon the charts" you wisely asked your cousin to outline the Gulf Stream on an Atlantic Ocean nautical chart so that "such chart and directions may be of use to our packets in shortening their voyages." The Franklin-Folger Chart, with its dark broad Gulf Stream, has notable concordance with our most refined understanding of Atlantic currents.

Publicizing the Gulf Stream enshrined you as a pioneering *oceanographer*. You went a step further, however, when you recorded the Gulf Stream's temperature variations during your transatlantic voyages. The tables of your three-times-a-day readings led to your conclusion that the thermometer may be "an useful instrument to a navigator," because it so readily identified the warm north-flowing current.

Today, sophisticated technologies provide sea captains with hourly images of sea currents; we obtain such pictures by measuring variations in seawater temperature. Thus, you were remarkably prescient when you predicted that the thermometer would be useful for charting currents.

By noting that one could also identify the Gulf Stream by the "gulf weed with which it is interspersed," you foresaw analyzing the stream by measuring the amount of the green plant pigment *chlorophyll* in the current and its surrounding waters.

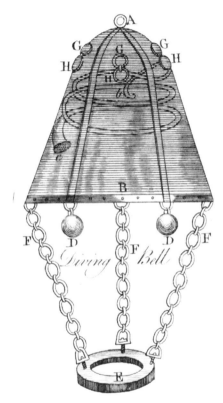

Finally, you recognized that the Gulf Stream could influence weather. You correctly surmised that the Gulf Stream's warm water would heat the overlying air, causing it to rise. The air on both sides of the Stream would be colder and thus heavier; "hence those airs must flow in to supply the place of the rising warm air, and meeting with each other, form those tornados and waterspouts frequently met with, and seen near and over the stream."

Further north, where the air surrounding the Gulf Stream is even colder, you rightly deduced, "the vapour from the gulph stream...when it comes into the cool air from Newfoundland, it is condensed into the fogs, for which those parts are so remarkable."

Do you recall hearing of a diving bell, an open-bottomed device employed in your own time for underwater salvage of sunken ships? It has evolved into completely sealed underwater ships—*submarines*—for both exploration and warfare.

On July 14, 1969, a Swiss-made submersible research vessel with a seven-man crew descended 2000 feet into the Atlantic Ocean for a month-long dive in the Gulf Stream. After its successful exploration, the submarine made other research dives before being retired. Today, that vessel is on display in a Canadian museum. Painted on its upper surface in bold black letters is the ship's name: *Ben Franklin.*

Perhaps you'll see it some day.

From: "Stuart Green" <stuartgreenmd@yahoo.com>
To: "Benjamin Franklin" <dr_benjamin_franklin@yahoo.com>
Subject: **Divide the hold of a great ship into a number of separate chambers**

Dear Doctor Franklin:

Oceanic science today includes oceanography (charting the ocean currents), undersea geology, marine biology, oceanic chemistry, and ocean technology and sea commerce (including marine architecture). In these next emails, I'll remind you of your contributions and insights to all these disciplines.

I find it easy to understand how you, spending so many hours at sea on your six transatlantic crossings, developed an interest in naval architecture, safety at sea, methods of propulsion and the like. Correctly realizing that you "may never have another occasion of writing on this subject" you decided to put your thoughts on maritime subjects into a lengthy 1785 letter to architect and ship designer Julien-David LeRoy: "I may as well now, once for all, empty my nautical budget, and give you all the thoughts that have, in my various long voyages, occurred to me relating to navigation."

Proposal for a 7 sail fore-and-aft arrangement on a single boom
From Franklin, *Experiments and Observations*, 4th edition

Fore-and-aft sail alignment, which you advocated, became the standard configuration for schooners years later. You mentioned that ship designers of your time, by "calculating to find the form of least resistance, seem to have considered a ship as a body moving through one fluid only, the water; and to have given little attention to the circumstance of her moving through another fluid, the air." You were referring here to the difficulty square-rigged ships had sailing towards the wind. You suggested that the *aerodynamics* (a self-evident word) of sails could be improved, "by dividing the sail into a number of parts, and placing those parts in a line one behind the other…"

Although your proposal for a slope with 7 triangular sails attached to a <u>single</u> boom never proved practical, by mid-19th century, multi-masted fore-and-aft cargo schooners plied the world's waters.

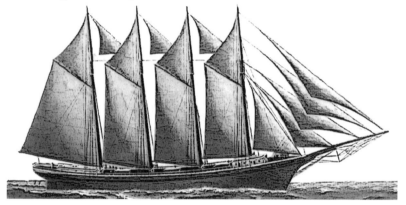

19th-century 4-masted fore-and-aft schooner

Lost anchors occurred with distressing frequency during outhaul from the sea. The anchor cable always snapped in a characteristic location—just where it attached at the anchor's hawser hole. You offered a practical solution for the problem: "To remedy this, methinks it would be well to have a kind of large pulley wheel, fixed in the hawse-hole, suppose of two feet diameter, over which the cable might pass; and being there bent gradually to the round of the wheel, would thereby be more equally strained, and better able to bear the jerk, which may save the anchor, and by that means in the course of the voyage may happen to save the ship."

Safety at sea was another issue you considered—and for good reason! You mentioned to LeRoy a dreadful thing that occasionally happened to ships: They are found on the high sea floating along "with no soul living on board, and so many feet of water in her hold," yet such vessels can be safely brought to port, indicating that the crew had abandoned ship. At other times, such empty ships wreck themselves against the shore.

In both cases, the crew and passengers left their vessels prematurely. You reported, "Those that give an account of quitting their vessels, generally say, that she sprung a leak, that they pumped for some time, that the water continued to rise upon them, and that despairing to save her, they had quitted her lest they should go down with her."

As you explained, water at first rushes rapidly into a leaking ship, "When a vessel springs a leak near her bottom, the water enters with all the force given by the weight of the column of water…This helps to terrify." Pumping out rapidly-entering water proves ineffective, you contended; the inflow slows, however, as the water level rises, making it possible for pumps to keep pace.

Here, as always, you made practical suggestions for keeping a leaking ship afloat. Empty water casks, you proposed, as well as sea chests containing light-weight materials, should be sealed tight ("bunged up" as you put it) and secured to low places in the hull to act as buoyant floats in the event of a leak: "if fixed, they help to support her, in proportion as they are specifically lighter than the water."

Watertight compartments—features borrowed from the East—are now incorporated in *all* ocean-going vessels. To remind you, Sir, here's how you put it: "one cannot help recollecting the well known practice of the Chinese, to divide the hold of a great ship into a number of separate chambers by partitions tight caulked…so that if a leak should spring in one of them the others are not affected by it."

While acknowledging "some little disadvantage" in ships with watertight chambers, you reasoned that the extra work involved in manning such a ship "might be more than compensated by an abatement in the insurance." Also, the added safety could be a selling point: "a higher price taken of passengers, who would rather prefer going in such a vessel." However, you expected your suggestions to fall on deaf ears because "our seafaring people are brave, despite danger, and reject such precautions of safety," fearing only the accusation of cowardice.

Capsizing at sea was an additional concern. You seemed impressed by the outrigger vessels of native seafarers: "The islanders in the great Pacific ocean, though they have no large ships, are the most expert boat-sailors in the world." You explained that Pacific islanders, when planning a long distance voyage, lashed two hulls together "by cross bars of wood that keep them at some distance from each other, and so render their oversetting next to impossible."

You also mentioned that Sir William Petty built a double-hulled vessel a hundred years earlier "to serve as a pacquet boat between England and Ireland." The ship made several crossings before being lost in a storm. You told LeRoy that "she needs no ballast, therefore swims either lighter or will carry more goods; and that passengers are not so much incommoded by her rolling: to which may be added, that if she is to defend herself by her cannon, they will probably have more effect, being kept more generally in a horizontal position, than those in common vessels."

Ever wanting to tinker with somebody else's ideas, you apparently couldn't accept Petty's design, which employed two normally configured hulls braced together. Instead, you suggested that each of the hulls have an unsymmetrical shape with the inner sides "perfectly parallel" to each other (*i.e.*, flat), and each outer surface be curved in the usual boat-shaped way.

Practical as usual, you even speculated about the potential cost constraints of future double hull fabrication: "The building of a double ship would indeed be more expensive in proportion to her burthen; and that perhaps is sufficient to discourage the method."

Today, many twin-hull ships incorporate your design. The cost of construction (as you surmised) restricts their use to ships built for speed and stability.

Fire at sea, a danger now as then, you observed, "is generally well guarded against by the prudent captain's strict orders against smoking between decks, or carrying a candle there out of a lanthorn [*lantern* today]." You also mentioned another practice that caused serious conflagrations at sea: "carrying store-spirits to sea in casks." Withdrawing alcohol from such casks near a burning candle caused at least two shipboard fires, one involving the *Serapis*, the British warship captured by your protégée John Paul Jones in 1779. You recommended that grog be stored in bottles only, thereby limiting the potential damage in case the beverage catches fire while being poured.

Lightning strikes at sea, you told LeRoy, can't happen with your lighting rod and chain apparatus, "now made and sold at a reasonable price by *Nairne and Co.*" The device is kept in a box and "run up and fixed in about five minutes, on the apparent approach of a thunder gust."

Collisions at sea, both with other ships and with icebergs, could be avoided with around-the-clock lookouts, you stated, a practice common in channels, "but at sea it has been neglected." Increased sea traffic and the islands of ice in the North Atlantic offered a good "reason for keeping a good look-out before, though far from any coast that may threaten danger."

Propulsion for ships also engaged your attention. You told LeRoy about a boat that crossed the Seine "in three minutes by rowing, not in the water, but in the air, that is, by whirling round a set of windmill vanes fixed to a horizontal axis."

We now use such air-propelled *blow-boats* in shallow swamps. After mentioning that a similar rotating wind vane had recently been placed on an air balloon, you went a step further with this noteworthy statement: "An instrument similar may be contrived to move a boat by turning under water" thereby anticipating the *propeller* by about a hundred years.

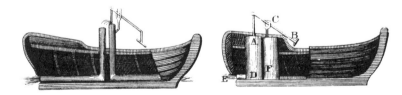

Your drawing of pump operated "jet boat"

You also considered ship propulsion systems that others had proposed. Most of these, you said, consisted of "paddles placed on the circumference of wheels to be turned constantly on each side of the vessel." Although such devices had been tried, they had "never been found so effectual as to encourage a continuance of the practice." You explained why: the paddles were set too deeply into the water, so that much of the effort in turning the drive crank was wasted pushing the water down or lifting it up. You thus foresaw the great *paddle wheelers* of a later era by suggesting that raising the paddle wheels "higher out of the water" would increase efficiency and avoided wasted effort.

You also anticipated the propulsion system of our modern personal watercraft, the *jet boats*. You repeated an idea first put forward by Bernoulli: construct on a boat an L-shaped tube with its vertical arm filled with water and its horizontal arm emptying out the "middle of the stern, but under the surface of the river." This will propel the boat forward, as long as water remained in the upper tube.

You recognized both the problem with such a system and its solution. Lifting water out of the sea with buckets to fill the upper arm of the tube would create drag in the water "that will be a deduction from the moving power." You proposed, therefore, that a second L-shaped tube with its underwater end projecting forward, "being worked as a pump, and sucking in the water at the head of the boat, would draw it forward while pushed in the same direction by the force at the stern." It would be necessary, of course, to calculate whether pumping the water through such a contraption required more or less effort "than that of rowing."

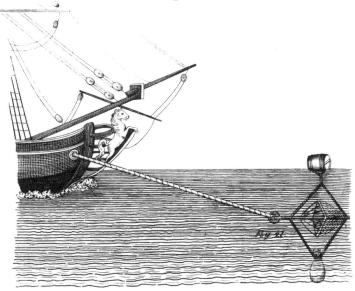

Your sea anchor proposal, from *E & O*

The last sentence of this proposal confirms you, once again, as a visionary: "A fire-engine might possibly in some cases be applied in this operation with advantage." You were referring to the *steam engine*, developed and fabricated by the firm Boulton & Watt. Steam engines, during your era, had but one practical application—pumping water out of mines. Your close relationship with Matthew Boulton and acquaintance with James Watt brought you in contact with their machine, which became the world's propulsive power in the century after your hibernation commenced.

Your fertile mind offered yet another proposal on ship propulsion—compressed air, a concept that to my knowledge has never proven effective. You suggested that a boatman could manually operate an air pump that exhausted below the waterline. By having two such devices—one for each hand—discharging air on either side of a vessel's stern, the boatman "might steer himself by working occasionally more or harder with either hand."

Hoving-to in a storm requires seamen to strike their sails and wait for the weather to pass. To prevent a ship from being blown towards a lee shore you described the construction of a "swimming anchor," which is now called a *sea anchor*.

You told LeRoy of a sea anchor shown you by "an ingenious old mariner." The device, about 25 feet long and a few inches wide, floated on the ocean's surface, which you thought a deficiency because it was "liable to be hove forward with every wave," reducing its effectiveness.

You suggested, instead, "two machines for this purpose," which resemble modern **sea anchors**: "The first is to be formed, and to be used in the water on almost the same principles with those of a paper kite used in the air. Only as the paper kite rises in the air, this is to descend in the water." You offered precise directions for an underwater canvas kite. Your sea anchor folded for storage by removing the canvas and rotating the spars until they paralleled one another.

You didn't stop, however, with one sea anchor design, offering a second as well. "The other machine for the same purpose," you explained, "is to be made more in the form of an umbrella." This design incorporated a square wood frame supporting the canvas around its edge.

Speaking of kites, you anticipated a sport we now call ***parasail-surfing*** by two hundred years when you wrote, "A man who can swim, may be aided in a long traverse by his handkerchief formed into a kite, by two cross sticks extending to the four corners; which being raised in the air, when the wind is fair and fresh, will tow him along while lying on his back, with the greatest pleasure imaginable"

In spite of the many proposals you offered for improving life at sea, you doubted seamen would listen to your advice: "Some sailors may think the writer has given himself unnecessary trouble in pretending to advise them; for they have a little repugnance to the advice of landmen, whom they esteem ignorant and incapable of giving any worth notice." You reminded LeRoy that "most of their instruments were the invention of landmen" including "the first vessel ever made to go on the water."

From: "Stuart Green" <stuartgreenmd@yahoo.com>
To: "Benjamin Franklin" <dr_benjamin_franklin@yahoo.com>
Subject: **One of Cook's vessels**

Dear Doctor Franklin:

By your time the great age of global discovery had already ended and a new period of information-gathering sea voyages was just beginning. Although you never joined a scientific expedition, you clearly supported such missions.

Do you recollect that while in London for the second time (to persuade the Penn family to pay their share of Pennsylvania taxes) you frequently spent Monday evenings with friends at the George and Vulture Restaurant? James Cook, a particularly bright young naval officer, often joined your group, which included some leading scientific thinkers of the time.

Cook obviously impressed his naval superiors with his intelligence and determination because, ten years later, the Royal Navy ordered Cook to lead a major expedition.

Edmund Halley, the celebrated astronomer, predicted that the planet Venus would cross in front of the Sun on June 3, 1769. By measuring the time it took for the planet's transit across the solar disc, astronomers could calculate both the Sun's size and the Earth's distance from it. The Royal Geographic Society recommended that the Navy send expeditions to three remote locations to obtain the necessary data—a south Pacific island, the top of Norway, and Hudson's Bay in Canada.

Cook got picked for the Pacific mission. He set sail on August 25, 1768, easing the 368-ton *Endeavour* out of Plymouth harbor. On board were eleven scientists, including your friend Sir Joseph Banks, who fifteen year later would correspond with you about ballooning and become President of the Royal Society.

Cook's journey on the *Endeavour* took him to Rio de Janeiro, around the tip of South America and then into the Pacific. He and his crew set up an observatory on Tahiti and made the appropriate celestial measurements.

After completing that part of the mission, Cook searched for a possible continent at the South Pole. Sea ice, however, kept him from sailing far enough south to discover anything. Turning west, Cook became the first European to visit what we today call *New Zealand*, which he later circumnavigated and claimed for the Crown. He next explored unknown parts of Australia and also added that region to England's possessions.

During the expedition, Cook insisted that his crewmembers eat fruit where available, vegetables and especially sauerkraut. They were thus among the first scurvy-free seamen in the Royal Navy, although many succumbed to illnesses they picked up in the tropics—including venereal diseases.

Cook's return to London in 1771, as you know, caused much celebration by the British whose empire he expanded. His description of the Maori people of New Zealand and his account of its flora and fauna resulted in an interesting proposal.

Alexander Dalrymple, an experienced Far East traveler, came up with the idea for a private expedition to New Zealand for humanitarian purposes. He solicited like-minded individuals, getting his greatest support from you, Dr. Franklin. Together with Dalrymple, you and other members of the group published the pamphlet *Introduction to a Plan for Benefiting the New Zealanders.*

Speaking of the native New Zealanders, you wrote, "The inhabitants of those countries, our fellow-men, have canoes only; not knowing Iron, they cannot build ships: They have little astronomy, and no knowledge of the compass to guide them; they cannot therefore come to us, or obtain any of our advantages." Proposing to bring the benefits of civilization to the natives, the pamphlet asks, "does not Providence, by these distinguishing favours, seem to call on us, to do something ourselves for the common interests of humanity?"

Acknowledging the less-than-honorable intent of many expeditions, your group wrote, "many voyages have been undertaken with views of profit or of plunder, or to gratify resentment; to procure some advantage to ourselves, or do some mischief to others."

Dalrymple, you and other supporters suggested sending the mission to New Zealand "not to cheat them, not to rob them, not to seize their lands, or enslave their persons; but merely to do them good, and enable them as far as in our power lies, to live as comfortably as ourselves."

To help sell the proposal, Dr. Franklin, your group suggested that England would benefit financially, because "a commercial nation particularly should wish for a general civilization of mankind, since trade is always carried on to much greater extent with people who have the arts and conveniences of life, than it can be with naked savages."

The group proved unable to raise the needed money, a disappointment especially for Dalrymple, who was deeply offended that Cook had been chosen to take the *Endeavor* south instead of him. (He nevertheless gained fame as an ocean mapmaker.)

After a year's sojourn in England, the Admiralty sent Cook on a second journey to find and claim for Britain any land at the South Pole. On board, Cook carried a new time-keeping device, the chronometer, claimed by its inventor John Harrison to maintain accuracy no matter how rough the seas.

Harrison's instrument, which went through four different models until it reached perfection, proved remarkably reliable, influencing the accuracy of all future measurements of longitude. (Do you recollect sending a book about "Harrison's Watch" to Debbie in 1767, for her to pass along to Philadelphia Watchmaker Edward Duffield?)

As with his first voyage, Cook couldn't penetrate the sea ice to find land near the South Pole. Moreover, he became convinced that no such land existed as he circumnavigated an ice mass that extended all the way across the Southern Ocean. (He proved wrong, Dr. Franklin: the continent *Antarctica* straddles the South Pole.)

Undaunted, Cook sailed north into the Pacific Ocean, which he explored for the next three years, becoming the first European to visit numerous islands in the region. He returned home in 1775 to a deteriorating situation with the North American colonies.

Cook officially retired after his second mission but within months volunteered for what proved to be his final voyage, a search for the Northwest Passage—a sea route to Asia above North America. He left England with two ships in the early summer of 1776, just as you and your fellow Americans began your own venture into history on a ship of state.

After a call at New Zealand, Cook headed east, discovering Christmas Island, the Cook Islands and a cluster of several beautiful islands he named after his friend, the Earl of Sandwich. Cook and his crew next headed north to fulfill the original purpose of the mission. As with his search for Antarctica, sea ice prevented Cook from making headway to the north, so he gave up and returned to The Sandwich Islands (now called *Hawaii*) for repairs.

VILLAGE OF KAAWALOA, ON KEALAKEKUA BAY,
WHERE CAPTAIN COOK WAS KILLED.

The native Hawaiians assumed Cook was a long awaited god and treated him accordingly. Cook and his crew, however, didn't spend much time enjoying the adulation because they had a mission to accomplish; they soon sailed to Alaska for another try at the Northwest Passage. A problem with his ship's mast caused Cook to make a fateful return Hawaii on February 13, 1779.

Around that time, English scientists became concerned that hostilities on the high seas between British and American warships might spill over and involve Cook's vessels as they crossed the Atlantic on their way home. Do you recall that they prevailed on you, our envoy to France, to ensure safe

269

passage for Captain Cook? You willingly obliged: your 1779 safe passage document *To All Captains and Commanders of American Armed Ships* remains one of your best known pronouncements.

James Cook

You informed your countrymen that Cook's expedition was "an undertaking truly laudable in itself, as the increase of geographical knowledge facilitates the communication between distant nations, in the exchange of useful products and manufactures, and the extension of arts, whereby the common enjoyments of human life are multiplied and augmented, and science of other kinds increased to the benefit of mankind in general."

If an American ship encountered one of Cook's vessels, you asked that American captains "not consider her as an enemy, nor suffer any plunder to be made of the effects contained in her, nor obstruct her immediate return to England..." Instead, they should provide "all the assistance in your power" if needed.

As it turned out, James Cook never had the opportunity to enjoy the hospitality of American seamen. When he returned to Hawaii to repair his mast, Cook was injured during a scuffle with some Hawaiians and died shortly thereafter.

Cook's demoralized crew, rather than complete their mission to find the Northwest Passage, returned home in August 1780.

Hawaii remained an independent nation until 1898, when the United States of America annexed it—to the chagrin of many natives. Sixty-one years later, however, when Hawaii became our fiftieth state, Hawaiians celebrated with great joy. To this day, Sir, Hawaii remains a tropical paradise—and a favorite vacation spot for Americans.

From: "Stuart Green" <stuartgreenmd@yahoo.com >
To: "Benjamin Franklin" <dr_benjamin_franklin@yahoo.com>
Subject: **Draw a boat in deep water**

Dear Doctor Franklin:

Most Americans know you as an electrical researcher and inventor of useful gadgets (like your "long arm" to retrieve books from high shelves), but few realize how many different scientific disciplines attracted your interest. Often, you posed and then answered questions by performing experiments with a simple apparatus. Your effort to determine the relationship of a boat's speed to the depth of water under keel illustrates the point.

In 1768, you wrote to Sir John Pringle reminding him about a peculiar observation you both made in Holland two years earlier. While traveling in a horse-drawn canal boat, you noticed the vessel's unusually slow speed, so you asked the boatman about it. He replied that the "water in the canal was low." You reminded Pringle that "On being again asked if it was so low as that the boat touched the muddy bottom; he said, no, not so low as that, but so low as to make it harder for the horse to draw the boat."

Obviously puzzled by the claim, you said "neither of us at first could conceive that if there was water enough for the boat to swim clear of the bottom, its being deeper would make any difference." The boatman, however, insisted that the water's depth influenced timetables. Wanting to understand why, you informed Pringle that you "determined to make an experiment of this when I should have convenient time and opportunity."

Upon returning to England, you queried Thames River boatmen about "whether they were sensible of any difference in rowing over shallow or deep water." They confirmed the Dutch boatman's assertion. Moreover, you realized that water depth in canals might impact future civil engineering plans. You informed Pringle, "As I did not recollect to have met with any mention of this matter in our philosophical books, and conceiving that if the difference should really be great, it might be an object of consideration in the many projects now on foot for digging new navigable canals…."

You first constructed a 14-foot long wooden trough, filling it nearly to the top with water. By trimming a plank to fit into the trough, you could alter the plank's depth by supporting it with wedges. Next, you fabricated a 6-inch long model boat to fit the trough: "When swimming, it drew one inch water."

The set-up of your experiment, Dr. Franklin, would impress any modern marine scientist: "To give motion to the boat, I fixed one end of a long silk thread to its bow, just even with the water's edge, the other end passed over a well-made brass pulley, of about an inch diameter, turning freely on a small axis; and a shilling was the weight. Then placing the boat at one end of the trough, the weight would draw it through the water to the other." In this way, you could apply a constant force to the model, irrespective of the water's depth.

Most people will chuckle when they learn how you collected data without a proper timepiece: "Not having a watch that shows seconds, in order to measure the time taken up by the boat in passing from end to end, I counted as fast as I could count to ten repeatedly, keeping an account of the number of tens on my fingers. And as much as possible to correct any little inequalities in my counting, I repeated the experiment a number of times at each depth of water, that I might take the medium."

At the end of the trials, you determined that the difference in boat-speed between the shallowest and deepest water was "somewhat more than one fifth [*i.e.*, 20%]." You extrapolated to the canal-boat situation and concluded that if "four men or horses would draw a boat in deep water four leagues in four hours, it would require five to draw the same boat in the same time as far in shallow water; or four would require five hours."

In a modern sense, Sir, you proposed a hypothesis, tested it experimentally, and immediately recognized the practical consequences. In fact, you even analyzed the cost associated with excavating deeper canals because you concluded your letter to Sir John by wondering "whether this difference is of consequence enough to justify a greater expence in deepening canals is a matter of calculation, which our ingenious engineers in that way will readily determine."

18th-century canal and lock

Indeed, Dr. Franklin, such cost-to-benefit calculations remain with us. In fact, our nation's capital contains a giant version of your wooden trough (David Taylor Model Basin), a 3200-foot long structure housing three water troughs—quite an improvement from your 14-foot long apparatus, wouldn't you say? The building contains shallow- and deep-water basins and a high-speed basin where engineers test future ship designs.

As a result of such research, modern ship designers have developed keel side-wings that improve a sailboat's handling. Other naval architects have created a bulbous protuberance for a ship's bow, which increases seaworthiness. When you get out of the hospital, Dr. Franklin, I'll show you these odd-looking appendages at Philadelphia's waterfront.

From: "Stuart Green" <stuartgreenmd@yahoo.com>
To: "Benjamin Franklin" <dr_benjamin_franklin@yahoo.com>
Subject: **Their waters are fresh quite to the sea**

Dear Doctor Franklin:

In this email, I'll discuss a subject of considerable interest to you—the salinity of oceans.

You evidently knew that differences in seawater saltiness had been measured in multiple locations. We now realize that seawater chemistry depends on many interacting factors. Evaporation removes pure water, increasing the sea's salt concentration. Rain has the opposite effect, diluting salinity. Rivers usually bring fresh water to the ocean, but many contain dissolved salts and solids that alter sea chemistry.

Polly Stevenson Hewson posed a number of questions in her letters about the sea, salt water and the tides that you patiently answered, thereby preserving for us your thoughts about these matters. You disagreed, for instance, with the common notion that "all rivers run into the sea." You warned Polly that whenever one holds an opinion not generally accepted, it's better to "propose our objections modestly" because, "we shall, tho' mistaken, deserve a censure less severe, than when we are both mistaken and insolent."

You correctly acknowledged that various rivers do, indeed, run to the sea, for example "the Amazones, and I think the Oranoko and the Missisipi." These rivers, you told Polly, clearly empty into the ocean: "The proof is, that their waters are fresh quite to the sea, and out to some distance from the land." You then asked rhetorically, whether the Thames, the Delaware or the rivers entering into Chesapeake Bay, "whose beds are filled with salt water to a considerable distance up from the sea," ever reach the ocean. Clearly, you had doubts.

"The common supply of rivers is from springs," you told Polly, "which draw their origin from rain that has soaked into the Earth." While a river runs, it constantly evaporates, which reduces its flow. Moreover, the amount of evaporation, "is greater or less in proportion to the surface exposed" so that deep narrow rivers loose less water to evaporation than broad shallow ones. You said that if a river runs into a lake with a large surface area "whereby its waters are spread so wide as that the evaporation is equal to the sum of all its springs, that lake will never overflow."

Next, you extended the evaporating lake concept to a river running to the sea that meets a twice-daily tidal flow. If the river forms a broad bed, the situation could be analogous to a large lake, where the evaporation rate equals the incoming flow, so that no river water reaches the ocean.

You suggested that even though it collects the outflow of "the great Rivers Sasquehanah, Potowmack, Rappahanock, York and James, besides a number of smaller streams each as big as the Thames," the Chesapeake Bay might be such a body of water that does not flow into the sea. You supported the conjecture by pointing out that the "Bay is salt quite up to Annapolis."

In a letter to your brother Peter, you speculated on the **origin** of oceanic salinity. You freely admitted that "some very great naturalists" held the opinion that the salt in the oceans came from the "dissolution of mineral or rock salt, which its waters happened to meet with." You also recognized that such an outlook assumed without proof that "all water was originally fresh." You disagreed with this view: "I own I am inclined to a different opinion, and rather think all the water on this globe was originally salt, and that the fresh water we find in springs and rivers, is the produce of distillation."

By "distillation" I assume you meant *evaporation*, as you explained in the subsequent sentence: "The sun raises the vapours from the sea, which form clouds, and fall in rain upon the land, and springs and rivers are formed of that rain." The rock salt found in mines resulted from evaporation of seawater, "and that of course the sea is now fresher than it was originally."

Interestingly, Dr. Franklin, you didn't offer any greater *proof* for your speculation than did the naturalists who believed that seawater was originally fresh, rather than salty. You did, however, have a coherent theory about how salt mines and other great salt deposits formed, which most logically explained the evolution of salt-flats.

"It is evident," you proclaimed, "from the quantities of sea-shells, and the bones and teeth of fishes found in high lands, that the sea has formerly covered them." At some point in the past, according to your conjecture, either the seas fell away from the land or, alternatively, the land "lifted up out of the water to their present height, by some internal mighty force, such as we still feel some remains of, when whole continents are moved by earthquakes." Regardless of how water left the land, "valleys among hills might be left filled with sea-water, which evaporating, and the fluid part drying away in a course of years, would leave the salt covering the bottom." Later, soil "from the neighbouring hills" covered the salt, so that it could only be found by digging into the ground, as miners do.

You also offered another possible process for salt accretion inside caves: heat from the earth's center—displayed during volcanic explosions—causing evaporation of seawater that had previously leaked into underground caverns. The evaporated water passes through the volcano's exit hole "while the salt remains, and by degrees, and continual accretion, becomes a great mass."

At the end of your letter to Peter, you explained how you came to your conclusion: "This is a fancy I had on visiting the salt-mines at Northwich, with my son."

That trip deep into the Earth clearly impressed you. Indeed, Sir, it appears to have influenced your theory of the Earth's formation, the subject of a previous email. For now, however, as I'm late for work, I must go.

From: "Stuart Green" <stuartgreenmd@yahoo.com>
To: "Benjamin Franklin" <dr_benjamin_franklin@yahoo.com>
Subject: **The effect of oil on water**

Dear Doctor Franklin:

Historians interested in your life know of your fascination with oil and water. They're familiar with your trick of calming a pond by secretly pouring oil onto its surface. Your 1773 letter to English physician William Brownrigg contains many interesting insights into the oil-water question. In this email, I'll discuss a few of them.

You informed him that you tried to smooth with oil the surface of a wind-agitated pond in Clapham but you didn't succeed at first because you poured oil on the pond's leeward edge, so wavelets pushed the oil to shore. On the windward side, however, "the oil, tho' not more than a tea spoonful, produced an instant calm, over a space several yards square, which spread amazingly, and extended itself gradually till it reached the lee side, making all that quarter of the pond, perhaps half an acre, as smooth as a looking glass."

When younger, you "read and smiled at Pliny's account of a practice among the seamen of his time, to still the waves in a storm by pouring oil into the sea: which he mentions, as well as the use of oil by the divers." You were initially skeptical about this claim by an ancient Roman writer, but later changed your mind. You told Brownrigg: "In 1757...I observed the wakes of two of the ships to be remarkably smooth, while all the others were ruffled by the wind, which blew fresh." You asked your ship's captain about the difference and were told that cooks of those two vessels had just emptied "their greasy water thro' the scuppers, which has greased the sides of those ships a little."

I can understand why you didn't like the captain's attitude, because "this answer he gave me with an air of some little contempt, as to a person ignorant of what every body else knew." His arrogance resolved you to "make some experiment of the effect of oil on water when I should have opportunity."

Before you conducted trials, you observed another curious effect of oil and water in vessels at sea. Your cabin lamp was a drinking glass suspended by a wire sling. Water partially filled the glass; illuminating oil and wick floated on the water.

As the ship rolled, you observed "the surface of the oil was perfectly tranquil, and duly preserved its position and distance with regard to the brim of the glass, the water under the oil was in great commotion, rising and falling in irregular waves, which continued during the whole evening." During the night, the flame consumed oil until only a thin coat remained. You noticed that the "the water was now quiet, and its surface as tranquil as that of the oil had been the evening before." When the oil was refilled, "the water resumed its irregular motions, rising in high waves almost to the surface of the oil, but without disturbing the smooth level of that surface."

Later, on land, you hung up a glass tumbler containing oil and water and started it swaying. Once again, you saw that the exposed oil surface remained calm, whereas the oil-water *interface* "was agitated with the same commotions as at sea."

You told Brownrigg that both our ancestors and primitive peoples made observations that were often ignored: "it has been of late too much the mode to slight the learning of the Ancients. The learned too, are apt to slight too much the knowledge of the vulgar."

18th-century cargo ship

You enumerated various ways people used oil on water in their daily activities. For example, placing oil on water was the "practice of the Bermudians when they would strike fish, which they could not see if the surface of the water was ruffled by the wind."

Likewise, "the fishermen of Lisbon when about to return into the river, (if they saw before them too great a surff upon the bar)" emptied a bottle or two of oil onto the water to "suppress the breakers and allow them to pass safely."

Lastly, "in the Mediterranean…the divers there, who when under water in their business, need light, which the curling of the surface interrupts, by the refractions of so many little waves, they let a small quantity of oil now and then out of their mouths, which rising to the surface smooths it, and permits the light to come down to them."

You added two more examples, both from commercial fishing. The Rhode Island whale harbor remained remarkably calm, a consequence of so much blubber dumping into the water; herring processing in a Scottish bay had an equally calming effect on waves.

In light of today's concepts, you had a reasonably coherent theory about why oil calmed the surface of water, consistent with your era's understanding of "affinities." You started out by noting that water has an affinity for air.

You next described how wind produced waves, an explanation as valid today as in your era. "Therefore air in motion," you said, "which is wind, in passing over the smooth surface of water, may rub, as it were, upon that surface, and raise it into wrinkles, which if the wind continues are the elements of future waves." The continued action of the wind, even a light one, will act on the small waves "rising higher and extending their bases, so as to include a vast mass of water in each wave, which in its motion acts with great violence."

Oil and water, if I correctly understand your view, repel each other; there also exists a "mutual repulsion between the particles of oil." Thus, "oil dropt on water will not be held together by adhesion to the spot whereon it falls, it will not be imbibed by the water, it will be at liberty to expand itself, and it will spread on a surface that besides being smooth to the most perfect degree of polish." You concluded that the oil layer kept the air and water apart, "and prevents friction as oil does between those parts of a machine that would otherwise rub hard together." Correct me if I'm wrong, Sir, but didn't you view oil as a lubricant, reducing friction between air and water? If so, you had a distinctly modern perspective on the subject.

It's obvious that you enjoyed a bit of showmanship with your oil-on-water demonstration. I know about your hollowed-out cane filled with oil that, in the presence of company, you'd pass above a wind-driven pond, much as Moses held his staff over the Red Sea. Your companions marveled at your ability to calm the waters, at least 'til you let 'em in on the joke.

After such a display, one witness—a Dutch sea captain—offered you a chance to test your concepts on the high seas. You told him about a problem Captain Cook encountered "particularly where accounts are given of pleasant and fertile islands which they much desired to land upon...but could not effect a landing thro' a violent surff breaking on the shore, which rendered it impracticable."

You suggested that pouring a sufficient quantity of oil on water to make a risky landing possible would justify the "the expence of oil that might be requisite for the purpose."

Shortly thereafter, near Portsmouth, you and the Dutch captain conducted an experiment using a longboat trying to beach in rough surf. As spectators watched, "The experiment had not in the main point the success we wished; for no material difference was observed in the height or force of the surff upon the shore."

Some observers did notice "a tract of smoothed water" headed towards shore, but overall, the experiment failed. Still, you weren't discouraged by the unexpected outcome: "It may be of use to relate the circumstances even of an experiment that does not succeed, since they may give hints of amendment in future trials."

You nevertheless tried to understand what happened. You concluded that oil couldn't calm seas already built to considerable height. Waves have their own inertia, "For we know that when wind ceases suddenly, the waves it has raised do not as suddenly subside, but settle gradually and are not quite down till long after the wind has ceased."

As for the "commotion" you noticed at the oil-water interface in a swaying lamp, you couldn't figure it out and declined to speculate. As you once told Sir John Pringle, you demonstrated the oil and water layer effect "to a number of ingenious persons" but nobody could explain why the oil's top surface remained flat while the oil-water boundary was in turmoil during rough seas.

Indeed, you correctly observed that "those who are but slightly acquainted with the principles of hydrostatics, &c. are apt to fancy immediately that they understand it, and readily attempt to explain; it but their explanations have been different, and to me not very intelligible. Others more deeply skilled in those principles, seem to wonder at it, and promise to consider it."

Do you remember the letter you received from London merchant Christopher Baldwin, a witness to your oil-on-water demonstration at the pond? He wrote during the peace negotiations ending the American War for Independence. Baldwin expressed the commonly held British view that you personally created the United States of America, set the course of war when Britain wouldn't agree to the independence of her former colonies, and then secured peace when the fighting ended: "Have you forgot your throwing oil on the pond near me—impossible! But a far, infinitely far greater object rises before me! 'Tis you my dear Sir who have troubled the mighty ocean. 'Tis you who have raised billow upon billow and called into action kings, princes and heros! And who after engaging the attention of every individual in Europe and America, have again poured the oil of peace on the troubled wave, and stilled the mighty storm"

From: "Stuart Green" <stuartgreenmd@yahoo.com>
To: "Benjamin Franklin" <dr_benjamin_franklin@yahoo.com>
Subject: **Propelling boats on the water, by power of steam**

Dear Doctor Franklin:

Steam power. No single term more completely defined the 19th century's *Industrial Revolution* than its predominant form of mechanical muscle—leading to steamboats, steam engines and steam-powered manufacturing. The seeds for such technology were sown in your era by innovative thinkers and mechanical designers, some well known to you. To support their efforts, investors bankrolled development corporations, hoping to reap the financial fruit of successful concepts.

Your 1788 Rumseyian Society of Philadelphia was one such venture, created to subsidize the steamboat design of James Rumsey, the Virginia inventor. He had, according to the charter, "been several years employed, with unremitted attention and at a great expence, in bringing to perfection the following machines and engines, namely, one for propelling boats on the water, by power of steam…"

Your group's first objective: send Rumsey to Birmingham to negotiate with Matthew Boulton and James Watt, patent holders of the most advanced steam engine available at the time. Your role in the endeavor, Dr. Franklin, was thought critical for success because you knew Boulton personally, having spent time with the English industrialist many years earlier.

Of course, not everyone in Philadelphia supported Rumsey's proposal. A rival group threw its weight behind that other steamboat pioneer, John Fitch. Unlike Rumsey, whose initial steamboats had too many technical problems to perform reliably, Fitch held a successful demonstration of his boat on the Delaware River on August 22, 1787. His astonished witnesses were members of the Constitutional Convention, in town to draft a new national governing document.

Because Fitch pestered George Washington, you, and other members of the Convention to grant a patent on his contraption, the delegates temporarily suspended their debates and ambled over to the Delaware River to watch the show. Fitch's boat, about forty-five feet long, had a primitive steam engine and Indian-style paddles (attached to reciprocating boards) that rowed the craft through the water.

In spite of Fitch's success that day, the principle delegates assumed that Rumsey had a prior claim for a steamboat patent, based on his own steamboat experiments on the Potomac River both before and after Fitch's presentation. The Fitch-Rumsey controversy raged on for five years. Fitch eventually viewed you, Dr. Franklin, as the leader of his opposition, comparing you to infamous villains in history who "made themselves great by fomenting wars, murder, rapine [now *rape*], and depopulating countries."

Although Hero of Alexandria, in the first century B.C., realized that expanding steam had the power to do work, no practical use of steam power appeared until the 18th century, and then only because advances in metal-working made possible boilers that could tolerate high internal pressure without blowing apart.

In 1698, British military engineer Thomas Savery came up with a functional application for steam power. His "engine" had no moving parts; instead, he used steam pressure from an enclosed boiler to force water up a pipe into a chamber, which was sprayed with cold water. The steam, as it condensed, created a vacuum pulling up even more water.

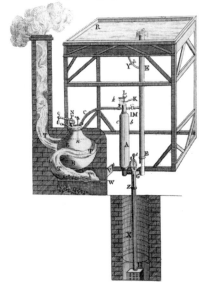

Savery Engine

Savery designed his pumps for mines but most were used to raise water to rooftop cisterns above the homes of wealthy patricians. In a 1702 advertisement for his device, Savery promised "such an engine may be made large enough to do the work required in employing eight, ten, fifteen, or twenty horses to be constantly maintained and kept for doing such a work."

Comparing the output of a steam apparatus to the power of horses became, thereafter, a standard measure for future steam engines.

For a period of time, Savery worked with a young ironmonger who later invented a steam-powered pump that revolutionized the way water was removed from mines. His name, you'll recall, was Thomas Newcomen, and he designed a machine that lasted to the early 20th century.

The Newcomen engine is considered the first *true* steam engine because, unlike Savery's, it had a vertically mounted *piston*. Steam entered the bottom of the cylinder when the piston was at the top of its stroke. Cold water sprayed over the cylinder cooled the steam, causing it to condense and create a vacuum within the cylinder. Atmospheric air entered the cylinder's top and pushed the piston downward and with it, the piston's rod. A large rocking arm attached to the piston rod provided the motion needed to lift water and do other tasks.

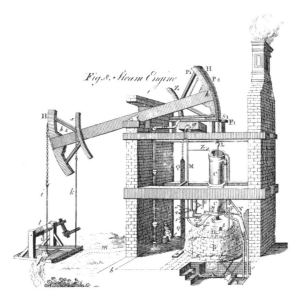

Newcomen engine

Newcomen engines soon became a common sight wherever miners tunneled below the water table. No doubt you saw them while traveling to England's mining regions in 1759 and 1771.

The next step in steam engine evolution involved a person whose name everyone today knows, James Watt of Glasgow, Scotland—a scientific and mathematical instrument maker who built equipment for the University of Glasgow. There, in 1756, he was asked to make instruments for Joseph Black, a professor at the University—and later a personal acquaintance of yours.

Black, as I mentioned previously, discovered carbon dioxide (called fixed air at the time) while doing research on the antacid Magnesia Alba. By the time he ordered equipment from Watt, however, Dr. Black had begun research on specific and latent heat for Scotland's whiskey industry. (By the mid-1700s, half the cost of producing distilled spirits went up in smoke.)

Black needed fine measuring instruments for the project; he asked Watt to make them.

At around the same time Watt was fabricating instruments and equipment for Joseph Black, he received for repair the university's model Newcomen engine. It had been fixed once, but failed to work properly. Watt, now understanding the concept of specific heat from his collaboration with Black, suspected a heat-related design flaw in the Newcomen machine.

Watt went for a Sunday stroll in May 1765, when the solution came to him. He later wrote, "I had not walked further than the golf-house when the whole thing was arranged in my mind." To increase the Newcomen engine's efficiency, Watt, instead of cooling the piston for each stroke, created a separate condenser for the steam. In that flash of insight, he increased the steam engine's efficiency four fold.

James Watt

Only in Scotland, Dr. Franklin, could a man thinking about whiskey while walking towards a golf house come up with an idea that changed the course of history.

After having created an improved steam engine on paper, making a working model required a substantial investment. Luckily, Watt teamed up with a successful British industrialist, your friend Matthew Boulton, to form a company whose steam engines became legendary for their efficiency and output—and high cost. More about this in my next email.

From: "Stuart Green" <stuartgreenmd@yahoo.com>
To: "Benjamin Franklin" <dr_benjamin_franklin@yahoo.com>
Subject: **A sooty crust on the bottom of the boiler**

Dear Doctor Franklin:

Matthew Boulton of Birmingham inherited a silver and metal works from his father and grew the business into a large manufacturer of buttons and coined money for the British government. Even before joining Watt in the steam engine venture, Boulton had hundreds of employees working in a mechanized factory powered by a waterwheel, connected via belts and pulleys to manufacturing equipment. He also had a Newcomen pump that caused him problems. Boulton tried unsuccessfully to improve the machine himself in 1766, making him receptive to Watt's proposal a couple of years later.

Matthew Boulton

Do you recall that, within a month of your House of Commons testimony to get the Stamp Act repealed, you received from Boulton a letter asking for guidance? He sent a model of his steam engine, telling you: "My engagements since Xmas have not permitted me to make any further progress with my fire engine," Boulton admitted, adding "but as the thirsty season is approaching apace, Necessity will oblige me to set about it in good earnest."

Boulton asked you two technical questions: "which of the steam valves do you like best;" and, "is it better to introduce the jet of cold water in at the bottoms of the receive (which is about feet from the top) or in at the top...?" Boulton, in soliciting your thoughts on this subject wrote, "if any thought occurs to your fertile genius which you think may be useful or preserve me from error in the execution of this engine you'll be so kind as to communicate it to me."

It appears you didn't know how best to cool the engine's component, but in reply offered this sagacious advice: "I know not which of the valves to give the preference to, nor whether it is best to introduce your jet of cold water above or below. Experiments will best decide in such cases."

Boulton also mentioned the undesirable accumulation of wood ash in the boiler's firebox. You already had experience with the insulating effect of ash, a problem you solved for the Pennsylvanian fireplace. You advised Boulton to fix the "grate in such a manner as to burn all your smoke." You also realized that unburned fuel "forms a sooty crust on the bottom of the boiler, which crust being not a good conductor of heat, and preventing flame and hot air coming into immediate contact with the vessel, lessens their effect in giving heat to the water."

Although Watt's steam engines started as mine pumps, they soon powered every phase of industrial activity. The Boulton and Watt products were large but surprisingly efficient and well made. (At the Kew Bridge Steam Museum such a pump constructed in 1820 for a waterworks continues to function today.)

John Rumsey, a carpenter and cabinetmaker, first came to the attention of George Washington shortly after the Revolutionary War while working on a house for the general in Virginia. Washington later helped Rumsey get backing for a steam-powered boat proposal. Rumsey made an impressive demonstration on December 3, 1787, at Shepherdstown in what's now *West Virginia*.

The town hasn't forgotten the event: Today, you can attend the James Rumsey Technical Institute, enjoy a meal at the Rumsey Tavern, visit the Rumsey Steamboat Museum, admire the Rumsey Monument or participate in the Annual Rumsey Regatta on the Potomac. Clearly, almost everyone in West Virginia believes that James Rumsey was "The Father of the Steamboat"—unless they descend from John Fitch.

Unlike the steady and stable Rumsey, John Fitch displayed periodic bouts of behavioral mood swings that today would get him certified as mad. He assumed the world was against him for his religious beliefs, such as they were. Fitch professed Deism and thus rejected the divinity of Jesus and the authority of Scripture. He filled his autobiography with invective statements about his landlords, partners, employees, and hen-pecking wife.

The matter of steamboat propulsion came up during your first encounter with Fitch. His initial sketches of a mechanism for moving a boat through the water showed a water jet system similar to the one you earlier described to LeRoy. The only difference between your proposal and Fitch's idea was the way the upper receptacle filled with water. Your system used human power whereas Fitch provided a steam-driven pump.

Fitch realized that he needed two things to have a truly successful business venture: a patent and financial backing. The best way to gain monetary support, Fitch believed, would be to impress the most progressive member of society—you—with his idea. He therefore wrote to you about the "subject of a steam boat." In a grandiose but perceptive statement, Fitch declared that his invention was a matter "of the first magnitude" that will impact "sea voyages as well as for inland navigation." Steamships, Fitch proclaimed, "would be able to make head against the most violent tempests, and thereby escape the danger of a lee shore." The ship's steam engine could also "free a leaky ship of her water," Fitch asserted. His concept was a "very simple easy and natural way by which the screws or paddles are turned to answer the purpose of oars."

Fitch's steam boat

All Fitch asked of you, it seems, was to "make an essay under your patronage, and have your friendly assistance in introducing another useful art into the world." You folded Fitch's request and filed it away with the notation, "John Fitch boat to go by a fire engine."

It seems you wanted Fitch to realize that you had proposed a water jet propulsion mechanism for sea-going craft several years earlier in your letter to Leroy, a communication also read before the American Philosophical Society. Therefore, you sent the following note to the Society's secretary,

Francis Hopkinson: "Please to permit Mr. Fitch to take an extract from that part of my nautical paper, which describes sundry methods of moving boats by forcing water or air out of their stern: and inform him if you can, at what time the paper was read in the Society."

In the twelve years between the Declaration of Independence and ratification of the U.S. Constitution, each state acted as its own patent issuing authority, without reciprocity between them or with other governments. Fitch, for instance, managed to get a steamboat patent from Virginia, but he made no progress with Pennsylvania. I assume you saw flaws in Fitch's design (or heard about problems from eyewitnesses to the launch) and questioned the young inventor about them.

Two weeks after the demonstration, Fitch wrote to you, obviously responding to issues raised in the face-to-face meeting. "In a conference that I had with your Excellency," Fitch wrote, "I heard you mention, that the Philosophical Society ought to be furnished with a model of a steam engine, and having completed one upon a small scale, I should be exceedingly happy, Sir, should it meet with your patronage, so far as to recommend the purchase of it to the Society." (In other words, if you want a model, pay for it.)

Ever confident of his invention, Fitch told you, "I am now morally sure, from experience that a vessel may be propelled to great advantage thro the water by means of a steam engine, and have undertaken the works upon a large scale, but am apprehensive that the money raised will be insufficient for the purpose." He next appealed for funds.

Clearly, Fitch failed to impress you with his ingenuity, perhaps because of his abrasive personality. You must have also realized that vessels using the inefficient Newcomen engine would suffer from the same problem miners experienced: a high cost-to-output ratio.

A correspondent asked you about Fitch's "curious experiments on the Delaware…concerning this new Invention…"

You replied almost immediately. Yes, you had heard about Fitch's steamboat and "never doubted that the force of steam properly applied might be sufficient to move a boat against the current in most rivers." However, you offered this caveat: "when I considered the first cost of such a machine as the fire engine, the necessity of its being accompanied constantly by a skilful engineer who would expect good wages, to work it, and repair it on occasion, and the room it would take up in the boat, I confess I have feared that the advantage would not be such as to bring the invention into use."

As you correctly predicted, Fitch's steamboats never attained profitability. Perhaps you and the other members of the Rumseyian Society realized the error in Fitch's engine choice and decided to try a different approach. I'll describe how in the next email, but for now, I'm heading up to bed.

From: "Stuart Green" <stuartgreenmd@yahoo.com>
To: "Benjamin Franklin" <dr_benjamin_franklin@yahoo.com>
Subject: **A large boat rowed by the force of steam**

Dear Doctor Franklin:

On with the James Rumsey story: With adequate financial backing and the successful Shepherdstown launch under his belt, Rumsey sailed for England in 1788 to negotiate with Boulton and Watt to use their engine on a boat. He carried your letter of introduction to Benjamin Vaughan, the same man you corresponded with about lead poisoning.

You advised Vaughan that Rumsey invented "an improvement of boilers for steam engines, whereby nine-tenths of the fuel may be saved; and the machine applied with convenience to the forcing boats against the stream of rivers." Without mentioning Fitch by name, you said "He apprehends that another mechanician of this country is endeavouring to deprive him of such advantage, by pretending a prior right to the invention." Telling Vaughan, "Mr. Rumsey is a person of good character and reputation in this country," you solicited his help.

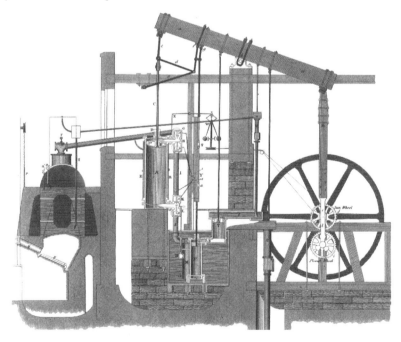

Watt and Boulton steam engine

By August 1788, Rumsey had met with Boulton and Watt in Birmingham and sent a progress report back to Philadelphia. His first encounter with the British industrialists proved less than satisfactory: Rumsey reported, "Messrs. Boulton & Watt" made "proposals that would have been dishonourable of me to except [accept]." With Vaughan acting as an intermediary through your "very philosophical letter," the negotiations improved.

"There is not a doubt but a large sum of money might be made in this country if I had the privilege of erecting boats to go by steam without interfering with Messrs. Bolton & Watt's patent privileges," Rumsey claimed, "Or could form a connection with them on reasonable terms."

Even though Rumsey never came to terms with Boulton and Watt, he started building the steamboat "Columbian Maid" based on his earlier propulsion system. Rumsey spent four years constructing the vessel, which made a successful public voyage on December 21, 1792. Rumsey, however, didn't see the launch. The prior evening, during a presentation about the boat to London's Society of Mechanic Arts, he suffered an apoplectic stroke. James Rumsey died the day his ship first steamed up the River Thames.

James Rumsey

Before Rumsey died, however, he met a young American, Robert Fulton, who had an interest in steam-powered boats—although he originally came to London to study art. Unlike Rumsey, Fulton did successfully negotiate with Boulton and Watt for an engine and set about to create a lucrative steamboat company, something yet to be accomplished anywhere.

In 1801, Fulton met the American ambassador to France, Robert Livingston (President Jefferson's appointment), who also fancied himself an inventor. A quarter century earlier, in 1776, Livingston worked with you, Jefferson, John Adams and Roger Sherman on the Committee that drafted the Declaration of Independence, so he was a Founding Father and thus had considerable influence back home.

When Fulton and Livingston returned to New York, they convinced the state's assembly to grant them an exclusive right to steamboat navigation on the Hudson River if they could produce a vessel that reliably traveled upstream from New York City to Albany at an average speed of 4 miles per hour.

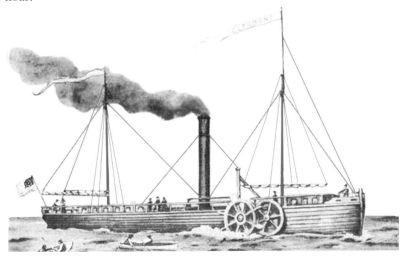

Fulton and Livingston's Clermont

After several years of frustrating failures —leading to the phrase "Fulton's Folly"—their *Clermont* made an impressive trip up the Hudson on August 7th, 1807, completing the 150-mile journey in 32 hours with an average speed of 5 miles/hr. The Fulton-Livingston team thereafter established the world's first profitable steamboat company.

As for John Fitch, the U.S. Congress finally granted him a patent on his steamboat in 1791, shortly after the Patent Act became law. To his chagrin, however, James Rumsey received a patent for his design on the same day, denying Fitch a potentially valuable priority claim on steamboats. Plagued by financial problems, Fitch finally gave up. In 1798, by then an alcoholic, Fitch gulped down a dozen opium pills and put an end to his high-pressure steam-filled life.

Steam powered boats became progressively more efficient. For transatlantic voyages, sailing ships added steam engines and paddle wheels amidships to provide motive power when becalmed. Eventually, square-riggers disappeared from the horizon.

At the beginning of the 19th century, inventors placed steam engines on carriages. At first, the devices couldn't move over uneven ground because of the boiler's weight. However, when steam powered carriages rolled on smooth, low-friction tracks, the problem was solved. Thereafter, *railroads* began to criss-cross the landscape.

Early British high-pressure steam locomotive invented by Richard Trevithick

The British led in steam engine and railroad technology, dominating with high quality products. By the middle and late 19th century and into modern times, the miles of railroad tracks within its borders usually reflected a nation's wealth, as powerful *steam locomotives* pulled thousands of tons of cargo across great distances.

In 1869, a transcontinental railroad joined the Atlantic and Pacific regions of the United States of America, bringing all inhabitants of this country to within a few days travel of each other.

Around the time engineers perfected the steam locomotive and steamship, inventors—mostly from continental Europe—developed the *internal combustion engine*. This machine uses a spark to ignite vaporized spirits, which pushes pistons within solid-walled cylinders as the hot gas expands.

In 1788, you wrote a letter to your French correspondent Jean-Baptiste LeRoy, informing the architect, "We have no philosophical news here at present, except that a large boat rowed by the force of steam now exercised upon our rivers, stems the current, and seems to promise being useful when the machinery can be more simplified, and the expence reduced."

That was a laudatory goal, Dr. Franklin, which engineers continue to pursue.

From: "Stuart Green" <stuartgreenmd@yahoo.com>
To: "Benjamin Franklin" <dr_benjamin_franklin@yahoo.com>
Subject: **The types he had been reading all his life**

Dear Doctor Franklin:

Any mention of Boulton or Watt must remind you of the Lunar Society of Birmingham. I know you enjoyed being a member of that like-minded group of natural philosophers and industrialists. They met on the night of a full moon to have bright illumination during their trip home; hence, the name. They focused attention on improving life through science and industrial modernization. The group—founded in 1765—eventually faded out of existence 23 years after your 1790 demise.

The Society's membership roster reads today like a *Who's Who* of prominent late 18th-century visionaries: Matthew Boulton, James Watt, Joseph Priestley, Erasmus Darwin, Josiah Wedgwood, William Withering, among others. I know that Anton Lavoisier and Thomas Jefferson corresponded with the group. And of course you attended meetings there.

I also suspect that you enjoyed visiting with Lunar Society member and type founder John Baskerville. His reputation as an outspoken agnostic caused problems for his next of kin after his death. Baskerville's religious beliefs (or should I say, lack thereof) precluded him from a churchyard burial, so he was interred on his estate. Later, when engineers dug a canal across the property, descendants exhumed him and secreted the body in a church crypt, only to be moved again to its present site—in a churchyard, interestingly enough!

I'm not surprised that church officials initially rejected his corpse in spite of his most famous printing job—a low cost Holy Bible. Making statements like this about revealed scripture will usually lead to charges of blasphemy: "I consider Revelation...exclusive of the scraps of morality casually intermixt with it, to be the most impudent abuse of common sense, which ever was invented to befool mankind."

Baskerville's agnosticism also led to attacks about his typeface by believers. I enjoyed reading how you defended him from one such assault. You informed Baskerville that an acquaintance told you that he didn't like Baskerville's type font, saying it was "a means of blinding all the readers in the nation, for the strokes of your letters being too thin and narrow, hurt the eye, and he could never read a line of them without pain."

You informed Baskerville: "In vain I endeavoured to support your character against the charge," but proved unsuccessful. So you decided to play a trick on the gentleman. You "tore off the top of Mr. Caslon's specimen, and produced it to him as yours brought with me from Birmingham, saying I

had been examining it since he spoke to me, and could not for my life per-
ceive the disproportion he mentioned, desiring him to point it out to me."

The man "declared that he could not then read the specimen without
feeling very strongly the pain he had mentioned to me." You never admitted the deception. As you put it: "I spared him that time the confusion of
being told, that these were the types he had been reading all his life with so
much ease to his eyes…nay, the very types his own book is printed with, for
he is himself an author; and yet never discovered this painful disproportion
in them, till he thought they were yours."

John Baskerville

To honor you friendship with Mr. Baskerville, all my emails are set in his
font, now available in every computer in common use.

PSYCHOLOGY & SOCIAL SCIENCE

✉ Varying degrees of imagination   295

✉ A convenient and handsome building   301

✉ A fine house…half a mile from Paris   305

✉ A good war or a bad peace   307

✉ Healthy and wealthy and wise   311

✉ May be bound in one volume…

with a complete index   313

✉ Disappointment and fallacious hope   319

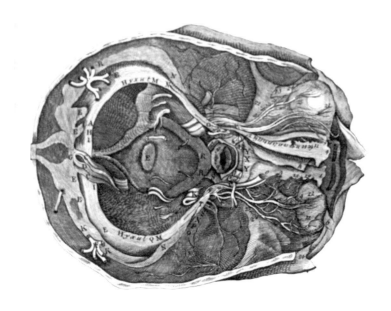

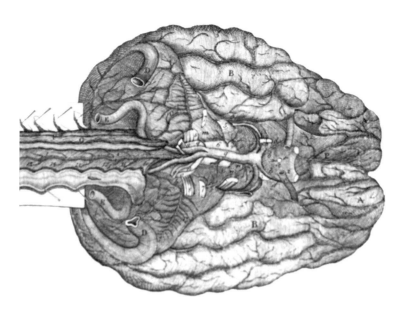

Interior of cranium and brain from Middleton's 1777 *Dictionary*

From: "Stuart Green" <stuartgreenmd@yahoo.com>>
To: "Benjamin Franklin" <dr_benjamin_franklin@yahoo.com>
Subject: **Varying degrees of imagination**

Dear Doctor Franklin:

Few people today understand the importance of the French Royal Commission's investigation of the mesmerism craze of the 1780s. But rest assured, Dr. Franklin, the commission's report guaranteed your place in the annals of clinical investigation. The strategy you and your co-commissioners employed to discredit Franz Mesmer's assertions about illness and treatment appear in the first published report of what we now call a *single-blind placebo-controlled clinical trial.*

The term *placebo* doesn't appear in Bailey's 1757 Dictionary, although in olden times it meant a flatterer or sycophant. At the beginning of the 19th century, the word came to designate a medicine used to placate a patient—a sugar pill.

The Report of the Royal Commission is rarely cited today. The late Stephen Jay Gould, a biologist, philosopher and apostle of rational thought, called your commission's report "...a key document in the history of human reason," primarily because of the method you and your fellow commissioners used to probe Mesmer's claims. Gould says the document "...should be rescued from its current obscurity, translated into all languages, and reprinted by organizations dedicated to the unmasking of quackery and the defense of rational thought."

In case you've forgotten details of the investigation, I'll briefly review the facts surrounding the phenomenon that literally *mesmerized* Parisians in France while you were there.

Franz Anton Mesmer, a Viennese physician arrived in Paris two years after you did, in February 1778. He was earlier forced out of Vienna by his medical colleagues who thought him a charlatan (and were perhaps jealous of the phenomenal growth of his practice).

Mesmer believed that an ultra fine magnetic fluid emanating from the stars and planets passed through all objects in the universe, both animate and inanimate. He attributed illness to a blockage of the fluid's natural flow and thought that he had the power to unplug the barrier.

At first, Mesmer used magnets for the curative procedure, sometimes after the patient swallowed iron filings. Subsequently, however, Mesmer dispensed with magnets and used his hands or an iron rod to affect the remedy. In this way, Mesmer emphasized the difference between treatments based on "mineral magnetism" that used magnets, and his version, "animal magnetism," which didn't.

The regimen started with pre-treatment preparation that utilized soft lighting, quiet music (often from the glass armonica you invented) and other measures known to evoke a trance-like state. Mesmer generated an altered type of consciousness in the patient. Part of the treatment involved inducing a "crisis" whereby a patient either fainted or convulsed uncontrollably.

On arriving in Paris, Mesmer, fearing resistance from other physicians, wanted to make a presentation to the learned medical society but was rebuffed. The public, however, quickly became enamored with Mesmer and his seemingly successful treatment, especially after members of the royal family—Marie Antoinette included—signaled their approval.

Franz Anton Mesmer

So many patients (and dilettantes) clamored for treatment that Mesmer and his disciples often used a "baquet" (bucket) for group therapy. The device—a large wooden tub usually about ten feet in diameter and one and a half feet deep—contained "magnetized" water and may also have held iron filings, broken bits of glass, or geometrically arranged bottles filled with water earlier magnetized by the treater. The baquet was covered by wooded planking through which protruded bent iron rods. The patients sat around the tub and either gripped a rod or touched it to the painful body part. Sometimes two or three circles of patients surrounded the baquet, interconnected either by silk ropes or by hold hands. (Do you notice the resemblance to electrical demonstrations?)

The mesmerist walked around the circle and touched the participants with his hand or an iron staff, inducing convulsions and fainting spells. When such sessions were held outdoors, a tree, properly "magnetized" by the mesmerist, substituted for the baquet.

Considering the dangers of traditional 18th-century medical treatment (with its purgatives to stimulate the bowels, emetics to induce vomiting, and bloodletting), it's no surprise that the alternative offered by Mesmer and his disciples proved popular. Moreover, Mesmer, being a medical doctor, had a knack for excluding from his clinic serious and incurable maladies and accepting mostly *psychosomatic* ailments (*i.e.*, those generated in the mind).

In 1784 Louis XVI, bowing to pressure from doctors to investigate mesmerism, created a Commission of Inquiry with seven distinguished physicians and scientists. (Did you realize that you were the first foreigner ever selected for a French Royal Commission?)

In case you've forgotten, other commission members included Antoine Lavoisier and Joseph-Ignace Guillotin, a well-known physician who argued successfully for "humane" executions.

Mesmer refused to participate in the commission's proposed plan for studying animal magnetism, so your group chose to scrutinize instead Dr. Charles d'Eslon, one of Mesmer's disciples.

From the outset, you and your co-commissioners decided to avoid looking into the mesmerists' alleged cures. Instead, you wisely focused on the validity of animal magnetism *per se* for, as your commission's report stated, "Animal magnetism may exist without being useful, but it cannot be useful if it does not exist."

I realize you had earlier given some thought to the mesmerism phenomenon. When a correspondent solicited your opinion, you astutely avoided the specifics of Mesmer's claims but wrote a statement that I include in all my lectures about the history of medicine: "There being so many disorders which cure themselves and such a disposition in mankind to deceive themselves and one another on these occasions...That delusion may however in some cases be of use while it lasts...If these people can be persuaded to forbear their drugs in expectation of being cured...they may possibly find good effects tho' they mistake the cause"

Recall that that Commission conducted investigations at several locations, including d'Eslon's clinic, Lavoisier's home and your Passy gardens. Gout and kidney stones prevented you from attending sessions conducted away from your residence. Your grandson Benjamin Franklin Bache, then fourteen, must have been astounded at what he witnessed during the experiments.

Both Benny's notes and the Commission's report describe how a boy was blindfolded (truly a "blind" assessment) and led to trees he was told had been magnetized by the mesmerist, although some were not. The lad responded to a placebo tree as though it were magnetized—he fainted and convulsed.

Similar inappropriate responses were observed at other test locations when the subjects were given a glass of water that either was or wasn't magnetized. Likewise, blindfolded patients convulsed when "magnetized" by fake mesmerists, while skeptical subjects (the commission members) couldn't be induced to respond, even by d'Eslon.

D'Eslon, after witnessing the outcome of the experiments, was persuaded by what he saw. According to your commission's report, he "declared in our session held at the house of Dr. Franklin the 19th of June [1784] that he thought he might lay it down as a fact that the imagination had the greatest share of the effects of animal magnetism..."

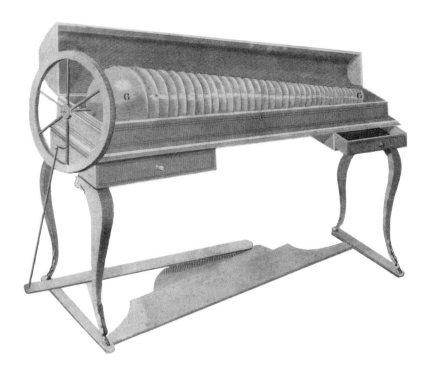

Your glass Armonica, used for background music during Mesmer's treatments

Your *Report of the Royal Commission* castigated mesmerism. The commissioners concluded that there was no such thing as animal magnetism, attributing the effects of mesmerism to the subjects' imagination. The report reads, "...only one cause is requisite to one effect, and that since the imagination is a sufficient cause, the supposition of the magnetic fluid is useless. In all of these experiments no differences were found other than those due to varying degrees of imagination."

Thousands of copies of the document were printed; the French viewed you as Mesmer's main antagonist. Indeed, the first English translation of the text was entitled, "Report of Dr. Benjamin Franklin and other commissioners..."

After the commission's report appeared, pamphlets criticizing or supporting the investigation followed. At least two stage-plays ridiculed mesmerism. A political cartoon showed a donkey-eared Mesmer, pockets stuffed with money, flying on a broomstick. Mesmer himself wrote you an angry letter claiming that d'Eslon wasn't a skilled practitioner of his method.

Some mesmerists charged that the commissioners unfairly condemned the greatest advance in health care ever discovered. They attributed the commissioners' motives to academic and professional despotism. These "radical mesmerists" produced documents that first attacked the Royal Commission report but gradually became vociferous assaults on the monarchy itself. The growing popularity of the radical mesmerists' perspective contributed, in some ways, to anti-royalist sentiment leading to the French Revolution.

I presume that in spite of your present memory difficulties, you'll remember the secret report you also presented to the King. It dealt with those aspects of mesmerism that the commissioners found salacious, such as the manner in which mesmerists stroked the lower abdomen and thighs of female patients while magnetizing them.

Mesmer later toured other French and European cities and finally settled down close to his birthplace in today's Switzerland. To the end of his days, Mesmer defended his concepts about animal magnetism and even drifted into occult practices. He became convinced that animal magnetism, properly applied, permitted him to see the future. He claimed to have magnetized the sun. Franz Mesmer died twenty-five years after your encaskment. Unlike you, Sir, he's not due back, although his name lives on in the word *mesmerize*.

The practice of therapeutic mesmerism evolved thereafter in several different ways. An Englishman named James Braid, realizing that a mesmerized subject was in a hyper-suggestible state, coined the term *hypnosis*. To this day, certain doctors employ hypnosis to treat irrational fears (*phobias*) and bad habits—overeating, tobacco smoking—but can't use it for addictions to opium or spirits.

Your Commission established a new strategy for assessing therapeutic effectiveness: the single-blind placebo-controlled clinical trial. In a *single-blind* study, the subjects of the experiment don't know whether they received active therapy or a sham treatment—the placebo. In a modern *double-blind* evaluation, both the subject and the experimenter are "blinded" to the intervention's identity.

Within a few decades came numerous placebo-controlled inquiries—both in Europe and America—into the professed benefits of many remedies.

Other blind assessments followed, scrutinizing everything from rheumatic fever to mental illness.

Nowadays, the double blind, placebo-controlled randomized clinical trial is considered the best way to evaluate the claimed value of newly developed pharmaceuticals and medical devices.

By being part of an investigation that pioneered the most reliable way to judge the efficacy of a wide variety of drugs and devices, you, Lavoisier and your fellow commissioners began a process leading to advances in medical therapeutics and the treatment of mental illnesses. Concurrently, blind assessment has put to rest many false treatment claims.

Lending your prestige and wisdom to the mesmerism inquiry undoubtedly contributed to project's success. Adding to A.R.J. Turgot's 1779 description of your principal achievements, it can truly be said of you: He snatched lightning from the heavens, the scepter from tyrants, *and the placebo from charlatans.*

A. R.J. Turgot

From: "Stuart Green" <stuartgreenmd@yahoo.com>
To: "Benjamin Franklin" <dr_benjamin_franklin@yahoo.com>
Subject: **A convenient and handsome building**

Dear Doctor Franklin:

Every year, a prominent news magazine lists America's best hospitals judged by reputation and other criteria. While many physicians and hospital managers—especially those at unranked facilities—dispute the placements, your and Thomas Bond's Pennsylvania Hospital remains one of America's finest.

Dr. Bond, you'll recall, was a Philadelphian who received his medical education in Europe, spending time in the London hospitals that provided both care of the poor and teaching patients for medical students. After establishing a successful practice in his hometown, Bond decided that a British style hospital would be good for Philadelphia, but he proved unable to get the project started; he wisely visited you to discuss the matter.

As you explained in your *Autobiography*, Bond "found there was no such thing as carrying a public spirited project through, without my being concerned in it." Bond told you that whenever he asked someone for support, they said, "Have you consulted Franklin upon this business? And what does he think of it?" Bond obviously made a compelling presentation because you were soon "engaged heartily in the design of procuring subscriptions from others."

I marvel at how you set out to "prepare the minds of the people by writing on the subject in the newspapers." Contributions flowed in for a while but then the pace slowed. At some point, it appears, you realized that the project needed legislative funding. As still happens today, your proposal to the colonial Assembly at first failed because representatives from rural areas balked at financing a hospital in Philadelphia.

The clever way you overcame their objections is what we now call a *matching grant* (you're often deemed its inventor) and is widely used to encourage donations for all manners of projects. By asking the Assembly to provide funds only if you could "raise £2000 by voluntary donations," the proposal passed. Members of the Colonial Assembly considered this a "most extravagant supposition and utterly impossible," so they enacted the requisite legislation "now convinced they might have the credit of being charitable without the expense."

Once this scheme was enacted into law, you had no trouble raising money "since each man's donation would be doubled." With the pledges in hand, the Assembly allocated the money and "A convenient and handsome building was soon erected…and flourishes to this day."

The tight control you and the other Managers maintained over the hospital resembled, in many respects, the prevailing British system. Many would now applaud your requirement that new doctors "give demonstrations of their skills and abilities in anatomy, operations, dressings and bandages..." Likewise, the Board of Managers approved each admission at weekly meetings.

While I recognize that the Board's activities occupied a considerable portion of your energy, you'd today agree it was time well spent, seeing how today's hospital staff treats you during confinement.

I compliment you, Dr. Franklin, on the way you raised money for the hospital again after the initial contribution request.

Modern fund-raisers should study your *Appeal for the Hospital* to learn from the acknowledged master of civic projects. At the time you wrote it, Philadelphia, a city of 15,000, was America's busiest seaport. Many foreigners poured in after lengthy sea voyages, some ill on arrival with no place to go for care. This fact gave you the first reason for establishing a hospital: "The increase of poor diseased foreigners and others" arriving in Pennsylvania needing treatment.

You warned prospective donors of risks to their own health by not supporting the venture.

Your *Appeal* also reminded potential contributors that their moral obligations "To visit the sick, to feed the hungry, to clothe the naked, and comfort the afflicted, are inseparable duties of Christian life." Since repetition never hurts and may prove helpful, you repeated the theological imperative: "History shows that from the earliest times of Christianity...publick funds and private charities have been appropriated to the building of hospitals, for receiving, supporting and curing those unhappy creatures, whose poverty is aggravated by the additional load of bodily pain."

Today's health officials employ a *cost: benefit analysis* to determine the value of in-hospital versus in-home treatment plans. Clearly, they should read what you wrote about the subject: "The difference between nursing and curing the sick in a hospital, and separately in private lodgings, with regard to the expense, is at least as ten to one." You proved your assertion by adding up all the expenses involved in a limb amputation and two months aftercare, totaling ten pounds all-inclusive. You concluded that, for the same outlay needed for one patient at home, *ten* patients could be cared for in a hospital.

More significantly, after acknowledging that the wealthy, when ill, can afford lodgings "that are commodious, clean and neat" and mentioning that the poor endure in "miserable loathsome holes" when sick, you asserted that "a beggar in a well regulated hospital, stands an equal chance with a prince in his palace, for a comfortable subsistence, and an expeditious and effectual cure of his diseases." (A remarkably prophetic statement, Dr. Franklin, considering the high quality health care provided in today's municipal hospitals.)

Philadelphia's tight-fisted Quakers, while rarely agreeing with you on military issues, opened their purses to support Pennsylvania Hospital for more than just diseased travelers and destitute patients. They also showed concern for the mentally ill.

To solicit money for the next phase of the hospital's expansion, you wrote and printed a handsome pamphlet, *Some Account of the Pennsylvania Hospital From its first Rise, to the Beginning of the Fifth Month, called May, 1754.* Here you mentioned the "inhabitants of the province, who unhappily become disordered in their senses, wondered about, to the terror of their neighbours, there being no place (except the house of correction) in which they might be confined, and subjected to proper management for their recovery…"

For these reasons, you said, Philadelphians chose to build a "hospital in the manner of several lately established in Great Britain" since you had earlier observed that "above two thirds of the mad people received into Bethlehem Hospital [London's 'Bedlam'], and there treated properly, have been perfectly cured."

Pennsylvania Hospital in the late 18ᵗʰ century

On May 28, 1755—a month after you supported General Braddock during the French and Indian War by supplying him wagons and communications assistance—you laid the cornerstone of the permanent Pennsylvania Hospital building.

For a while, it became nearly impossible for the hospital to stay solvent once the new facility's doors opened. In modern parlance, the hospital admitted too many "no-pay-patients." Fortunately, the Pennsylvania legislators, impressed with the hospital's handsome new building, voted £3000 to support the institution and its mission. Things settled down after that.

Many of Philadelphia's physicians joined the staff; medical student teaching commenced at Pennsylvania Hospital after 1765 when the College of Philadelphia created the nation's first medical school. For years thereafter, the hospital served as the teaching center for the medical school's faculty.

In 1771, your hospital co-founder sent you a Philadelphia *update*: "The hospital flourishes much, the old managers stick to it steadily," Dr. Bond said, adding (with obvious pride) that Philadelphia's new medical school was four years in operation: "The college is as when you left it, The School of Physic [Medicine]…could soon make a considerable figure, every branch of medicine is really well taught in it."

Changes at Pennsylvania Hospital during the two hundred years since your demise mirror developments in similar facilities across the country and throughout the world. Surgery during the time of Thomas Bond's tenure consisted of lithotomies, limb amputations, drainage of abscesses, and a few other quick operations. Large tumors might be removed if accessible.

Discoveries in the 19th century caused a revolution in hospital-based treatment. As I mentioned in prior emails, ether anesthesia and the evolution of bacteria-free surgery have made operations safer than ever. The 1895 discovery of X-rays brought hospital-based care into the modern era.

Your dedication to civic goals and your willingness to jump in with both feet, so to speak, to help push a project along, have endeared you to all, both during your times and ours.

In your *Autobiography* you attributed your penchant for good deeds to two books that, you said, "gave me a turn of thinking that had an influence on some of the principal future events of my life." One volume was Daniel Defoe's *Essay on Projects* and the other, *Essays to do Good*, written by your brother James's nemesis, Cotton Mather.

It was most gracious of you to later inform Samuel Mather, the elder minister's son, that *Essays to do Good* influenced your "conduct through life." Moreover, you thoughtfully wrote, "I have always set a greater value on the character of a *doer of good* that no any other kind of reputation." Young Mather must have certainly enjoyed reading, "if I have been, as you seem to think, a useful citizen, the public owes the advantage of it to that book."

It doesn't surprise me that you took particular pride in the way your matching grant idea helped launch Pennsylvania Hospital. I sensed glee in your tone when, in your *Autobiography*, you wrote, "I do not remember any of my political maneuvers, the success of which gave me at the time more pleasure. Or that in after-thinking of it, I more easily excused my-self for having made some use of cunning."

From: "Stuart Green" <stuartgreenmd@yahoo.com>
To: "Benjamin Franklin" <dr_benjamin_franklin@yahoo.com>
Subject: **A fine house…half a mile from Paris**

Dear Doctor Franklin:

Do you remember the house you lived in while Ambassador to France? You described it to Mrs. Stevenson thus: "a fine house, situated in a neat village, on high ground, half a mile from Paris, with a large garden to walk in." That structure was razed. Paris has expanded and Passy is now a district of the city. Perhaps one day you'll enjoy strolling down Rue Benjamin Franklin, looking for familiar vistas.

Mrs. Stevenson's residence on Craven Street in London, on the other hand, remains preserved today because you boarded there for so many years. The structure has recently been repaired. The committee in charge of restoration will soon solicit your advice about placement of furniture and other items.

Market Street, Philadelphia, today

Once you start to feel like yourself again, I'm sure you'll ask about your impressive 10-room house on Market Street, hoping to live there after you leave the hospital. I'm sorry to say, it was torn down in 1812. Only the foundation, the well and the privy pits remain.

Imagine, Sir, your privy pits!

Not only that, but pieces of the broken pottery you and your family threw into those holes has been retrieved by *archeologists* and is proudly displayed by the property's present owner, the U.S. government. Beneath the court-yard sits a fine museum about your life, a popular destination for Philadel-phia's tourists. Today, visitors to "Franklin Court" see two artfully con-structed metal frames designating the former location of your house and print shop.

Your print shop's location framed in Franklin Court, Philadelphia

With the money you'll earn from namesake royalties, speaking fees and publisher's advances, you needn't worry about finances; you'll be able to afford a fine Pennsylvania home as soon as you're ready to settle down somewhere beyond the hospital's walls.

Today in Philadelphia one can visit Franklin Court, walk past the Frank-lin National Bank, ambulate along Benjamin Franklin Highway, travel past Franklin Plaza, and enjoy the Franklin Institute—a science museum housing a colossal statue of you larger in size than the temple deity icons of ancient Greece.

So don't worry, Dr. Franklin, you'll be welcome wherever you chose to live in the City of Brotherly Love.

From: "Stuart Green" <stuartgreenmd@yahoo.com>
To: "Benjamin Franklin" <dr_benjamin_franklin@yahoo.com>
Subject: **A good war or a bad peace**

Dear Doctor Franklin:

While my emails have focused primarily on scientific matters, I assume you're curious about the course of world history during your two-century encaskment. Since your life spanned almost the entire 18th century—with you as its most eminent citizen—I'll describe, in a few broad strokes the principal events of the two centuries following your announced death.

A seminal occurrence in European history—the one most influencing further developments—started during your own lifetime. In fact, thoughtful historians hold <u>you</u> partially responsible for what we now call the *French Revolution*. You certainly understood, during your time in France, how to take advantage of Louis XVI's desire to avenge his Seven Years' War loss to Britain. The loans you extracted from France (about 5 million livres) bankrupted her. Additionally and more subtly, your appeal to French commoners—a leather-apron man who rose through effort and genius—stirred their republican dreams.

A complicated sequence of events eventually brought both Louis XVI and Marie Antoinette to the "humane" decapitation machine promoted by your friend Joseph-Ignace Guillotin, but this revolution created neither contentment nor prosperity for the people. A *Reign of Terror* followed, Dr. Franklin, in which tens of thousands died by the guillotine, including many of your acquaintances. For example, your friend Antoine Lavoisier—a descendant of tax collectors—lost his head in the Terror.

A Corsican military officer, Napoleon Bonaparte, slowly restored French pride and power with successful military campaigns that gained for France control over many of her neighbors. He gradually assumed for himself both the mantle and regalia of power. As Bonaparte conquered Europe, he placed relatives on thrones across the continent.

To finance an unsuccessful plan to invade England, Bonaparte sold France's North America possessions to America in 1803 (during Thomas Jefferson's presidency), doubling our country's size.

Napoleon's ultimate defeat by the combined forces of many nations brought a measure of tranquility to Europe.

In South America, former Spanish and Portuguese colonies—inspired by the American Revolution—rebelled against their European masters, freeing themselves from foreign domination. New Spain became independent *Mexico*, but soon lost its northern regions to the United States through warfare. Additional purchases and acquisitions completed America's transcontinental growth.

During the middle1800s, Europeans and their colonists (including Americans) completed their dominance of native peoples throughout the world. White people swept westward across North America and Australia and moved from south to north over much of Africa, India, and the region between India and China—now called *Indochina*. The indigenous peoples of these regions were either killed or subjugated or squeezed onto reservations and denied citizenship in their own lands. Britain, France, Belgium, Holland and other European countries competed for many of these territories, especially in Africa.

In the United States, a bloody Civil War over slavery (1861-1865) nearly split the nation in half. The northern (anti-slavery) states prevailed, and the abhorrent practice finally disappeared from our land.

Meanwhile, individual principalities in certain regions of Europe consolidated into nations, based primarily on shared language. Thus, Prussia came to dominate and unite the Germanic states, while a similar thing happened in Italy. These newly formed countries came late to the empire building game and thus had to catch up as best they could, sometimes leading to armed conflicts. Several wars broke out between neighboring European countries as they tried to acquire regions adjacent to both.

The 19th century, according to most historians, encompassed more than one hundred years. They say that it began on July 14, 1789, when French citizens attacked the Bastille prison and ended on June 28, 1914, when a Serbian nationalist assassinated the Austro-Hungarian Empire's heir-apparent. (The murder evolved from a religious conflict between the ruling Catholics and the subjected Serbian Orthodox Christians.) Thus the "Long Century," which lasted from your own lifetime to the modern era, started and ended with violence.

Great technological changes occurred during the 19th century that improved people's lives. In your time, the fastest message transmission occurred on the back of a galloping horse. It took six days, I've read, for news of Lexington and Concord to reach the Continental Congress in Philadelphia. By the end of the Long Century, messages traversed vast distances with the speed of your electric fluid, carried along copper wires. Likewise, electric lighting replaced fire for indoor illumination. Around the same time, as I've mentioned in prior emails, steam power—and later, internal combustion engines—displaced horses and oxen as motive force for land transportation.

Not all 19th-century inventions advanced man's happiness. Weapons of war became more deadly. Instead of reloading after each round, gun makers fabricated rifles holding many bullets, each with its own built-in charge. The killing efficiency of such firearms turned battlefields into slaughterhouses.

Employing Bernoulli's principle, as described earlier, fixed wing airplanes supplanted balloons for both air travel and warfare.

Aerial bombardment, prophesied in the letter Jan Ingenhousz wrote you about balloons, came to pass, wrecking great havoc during bomb attacks. Rather than reconsidering the "folly of wars," which you suggested might result from such weapons, national leaders relied on their destructive power for foreign policy purposes.

Patent application for 1862 Gatling Gun, a rotating cannon that revolutionized warfare

The 20th century saw cataclysmic upheavals as mighty nations fought each other over issues hard to comprehend in the light of history. As you once put it, "The flames of war, once kindled, often spread far and wide, and the mischief is infinite." The assassination of Austria's archduke triggered a devastating "World War" (1914-1918). The victorious allies, France, Britain, Russia, and the United States eventually dismantled the empires of the losers—Austria-Hungary, Germany and Ottoman Turkey.

Seething resentment in the defeated countries (especially Germany) percolated for nearly thirty years, finally erupting in a Second World War (1938-1945) involving much the same combatants—with Italy and Japan this time allying with the Germans. Again, the United States fought on the winning side and helped liberate the foreign possessions of the vanquished countries.

Despite retaining her dominions during the two great wars, Britain lost most of her colonial empire in the second half of the 20th century through liberation movements by local peoples. The same happened to other imperial powers as well.

Revolutionaries overthrew the Russian Tsar and nobility in 1917 hoping to create a utopian society. Within a few years, however, despots came to power and mercilessly repressed that region's people. The 20th century witnessed the growth and collapse of Russia's influence on smaller nations along her borders. About 15 years ago, that regime crumbled under its own corrupt weight, replaced by leaders professing democratic values. Imperial China followed a similar trajectory.

The 20th century ended, for all intents and purposes, on September 11, 2001. On that day, fanatic Mussulmen (now called *Muslims*) launched a new Holy War against the infidel West by killing thousands of New Yorkers in a surprise attack. Thus opened a new chapter in the sorry history of mankind since your demise.

Around the globe, extremist Muslims are increasingly hostile towards groups holding religious views differing from their own. They battle with Hindoos (now spelt *Hindus*) in India, Buddhists in Siam (now *Thailand*), Jews in Jerusalem, Christians in Europe and Asia, and amongst other Muslim sects in their own lands.

Nobody dares predict how the present century will end. I realize that you famously said: "I have been apt to think that there has never been or ever will be any such thing as a good war or a bad peace." I fear, Dr. Franklin, a bad peace descending upon us, with no end in sight as long as people remain convinced of the rightness of their own creed and the heresy of beliefs held by others.

You once said, "people increase and multiply in proportion as the means and facility of gaining a livelihood increase." Many of today's conflicts, in truth, stem from competition for increasingly scarce resources by an exploding world population—now more than 6.6 billion. This represents an eightfold increase from your time to now. The United States, whose growth rate you accurately predicted, is now the world's third most populous country after China and India. The increase followed control of the infectious distempers that carried off so many infants and children in your era.

Today, nations clash over agricultural land, water, precious and useful metals, and most significantly petræ-oleum (liquid bitumen, now called *petroleum*), a major source of heat and power, having supplanted both tallow and whale oil. Nobody, however, appreciated its great value during the 18th century when last you lived. Petroleum is *non-renewable*, meaning that once we deplete the supply, darkness may descend on humankind unless we find a suitable substitute.

To make matters worse, after you revitalize, millions will want to duplicate your Madeira-soaked/oaken-barreled experiment, hoping to hibernate for two hundred years, awaiting cures for today's hopeless diseases.

Within hours of your emergence, Sir, Madeira will disappear from wine merchants' shelves. Soon thereafter, all wine fermentation will cease—except for that producing Madeira. Grapes will become precious commodities; speculators will bid-up vineyards until only the wealthiest own them.

(You put it best when you informed David Hume, "The various value of every thing in every part of this world, arises, you know, from the various proportions of the quantity to the demand.")

As for oak, the world's forests will be stripped bare in a week. Hording will follow. Unscrupulous lumbermen will bleach other hardwoods, selling counterfeit wood as life-sustaining sheaves. This, in turn, will result in new legislation—The Oak-Barrel Laws—designed to stabilize the situation, but the demand will continue unabated. After all, who wouldn't spend their last pence for hope of future revivification?

With such a frightful prophecy, perhaps it would be better if my assumption about your *mort-faux* ultimately proves erroneous.

From: "Stuart Green" <stuartgreenmd@yahoo.com>
To: "Benjamin Franklin" <dr_benjamin_franklin@yahoo.com>
Subject: **Healthy and wealthy and wise**

Dear Doctor Franklin:

My email about steam power reminded me about something you'd enjoy. A renowned writer called Mark Twain wrote a humorous article about the misery your sagacious maxims inflicted upon young boys after your *embarilment-vivant*.

The Missouri-born author's real name was Samuel L. Clemens (1835-1910) and, like you, he started his literary career as a newspaper writer. Before that, he worked for a while on a river steamboat, one with a paddle wheel virtually identical to your proposal to LeRoy. Clemens took his *nom de plume* from the call of riverboat men when their bottom-sounding line reached the two-fathom marker.

Mark Twain

Clemens wrote that your "maxims were full of animosity toward boys." One of your most famous sayings, "early to bed and early to rise makes a man healthy and wealthy and wise" has caused untold suffering to boys everywhere, according to Clemens. Because his parents tried to make him follow such advice, "The legitimate result is my present state of general debility, indigence, and mental aberration." If his parents had just let him sleep long enough to "let me take my natural rest," he would likely have ended up a storekeeper, "respected by all" instead of what he became.

He said of you: "With a malevolence which is without parallel in history, he would work all day and then sit up nights and let on to be studying algebra by the light of a smouldering fire, so that all other boys might have to do that also or else have Benjamin Franklin thrown up to them."

By flying your lightning kite on a Sunday, Clemens claimed you were a "hoary Sabbath-breaker."

After many other complaints, Clemens, as loved by the public in his time as you were in yours, concluded his piece by lamenting his fate: "When I was a child I had to boil soap, notwithstanding my father was wealthy, and I had to get up early and study geometry at breakfast, and peddle my own poetry, and do everything just as Franklin did, in the solemn hope that I would be a Franklin some day. And here I am."

In a few days, I'll travel to Philadelphia to judge for myself the excavation's progress. I fear the worst: that the new National Constitution Center had been completed and you weren't discovered in the site.

From: "Stuart Green" <stuartgreenmd@yahoo.com>
To: "Benjamin Franklin" <dr_benjamin_franklin@yahoo.com>
Subject: **May be bound in one volume...with a complete index**

Dear Doctor Franklin:

Assuming that I've not wasted my time writing these emails, I thought that you might want to organize them in a box or binder, as you suggested with the *General Magazine and Historical Chronicle*. You advised your readers that the issues might be "bound in one volume...with a complete index" provide at years' end, so too have I prepared for you an index to help locate subjects of particular interest to you. Footnotes, citations and suggested readings are at the Friends of Franklin website, friendsoffranklin.org.

Here's the index:

A Guess at the Cause of Heat of the Blood in Heath and of the hot and cold fits of some Fevers, 179
acrylic plastics, 26
Adams, John, 27, 57, 143, 289
Adams, John Quincy, 27
Aepinus, 110
Africa, 308
airplane, 230
Albany Plan of Union, 19, 27
alchemists, 240
alchemy, 120
amber, 82, 83, 97
American Philosophical Society, 9, 26, 185, 206, 214, 235, 248, 285
Ampère, Andre-Marie, 97
An Account of the New-invented Pennsylvanian Fire-Places, 251
An Essay on the Principle of Population, 221
analytic chemistry, 127
animal magnetism, 141, 295, 297-99
Annapolis, 10, 93, 197, 274
anthrax, 50
antibiotics, 50, 70
Antoinette, Marie, 296
aortic and pulmonary valves, 190
apoplexy, 188
Appeal for the Hospital, 302
Aristotle, 175
Articles of Confederation, 27
asbestos, 200
atmospheric balloons, 227
atomic theory, 146
Aurora Borealis, 5, 223-6
Autumn Crocus, 34, 36
Bache, Benjamin Franklin, 14, 15, 297
Bache, Richard, 169

Bacon, Francis, 11, 157
bacteria, 48- 51, 59-6, 70, 75, 187, 304
bagatelle, 33
Baldwin, Christopher, 278
Banks, Joseph, 228, 229, 267
baquet, 297
Barbeu-DuBourg, Jacque, 9, 11, 13, 256
barometer, 233-5, 247-8
Baskerville, John, 291, 292
Bastille, 308
Beaumont, William, 160
Belcher, Jonathan, 100
Bell, Alexander Graham, 111
Bernoulli, Daniel, 230, 265, 308
Bessemer, Henry, 165
Bible, 195, 203, 253, 291
bifocals, 25, 26, 25–26
bioluminescence, 113
Bishop Ussher, 203
Bjerknes, Vilhelm, 242
Black, Joseph, 126-7, 281
bladder stones. *See* stones, bladder
Blanchard, Jean Pierre, 230
bleeding, 76
blood clotting, 188
bloodletting, 67, 77, 148, 297
boiler, 284
boils, 49
Bonaparte, Napoleon, 307
Bond, Thomas, 248, 301, 304
Boulton, Matthew, 266, 279-91
Bowdoin, James, 113-4, 212-4, 217
Boyle, Robert, 123, 124, 233
Braddock, General Edward, 303
Breintnal, Joseph, 250
Brillion, Mme, 42
British Museum, 200

Brownrigg William, 275
bubonic plague, 61
Buchanan, Andrew, 187
Buffon, Comte de, 88, 172, 202-3
Burr, Arron Sr., 100
Cabbeo, Niccolo, 82
Calvinism, 253
calx, 121, 125, 135-6, 138-9, 142
Cambridge, 39, 219
camp fever, 61
canal boat, 271
canal depth experiment, 271–72
carbon arc, 111
Caslon, 291
catching colds, 71
catheters, 29–32
Cavendish, Henry, 93-5, 128-34, 139-42,
    211, 227-31
central heating, 252
Cerletti, Ugo, 102
changing poles, 212
Charles, Jacques, 227
*Chemical Observations and Experiments on Air
    and Fire*, 33
chemistry, 119–22
chess, 42
Chinese, 52, 65, 71
cholera, 58, 196
Christ Church, 14, 81
chronometer, 268
Cinchona tree, 64
City Tavern, 163
Civil War, 308
Clemens, Samuel L.. *See* Twain, Mark
Clermont, 289
clinical electrotherapy, 102
clouds, 240
Cohen, I. B., 252
Cohn, Ellen, 10
colchicine, 36, 69
Colden, Cadwallader, 124, 179-191
Collinson, Peter, 84-92, 243-4
colors, 250
Columbia University, 64
common cold, 71, 75
Condorcet, Marquis de, 221, 234
Constitution Center, 312
Constitutional Convention, 38, 279
contagious diseases, 58
Continental Congress, 49, 141, 308
continental drift, 201
Cook, James, 6, 132, 257-269, 270, 277
Copley Medal, 89, 91, 104, 184
cost: benefit analysis, 302
cowpox, 53, 54
Craven Street, 183, 185, 305
Croghan, George, 199
Cullen, William, 73, 236
Cusa, Nicholas de, 235

d'Eslon, Charles, 297-9
Dalrymple, Alexander, 268
Dalton, John, 146-7
Darwin, Charles, 218-22
Darwin, Erasmus, 172, 219, 291
Darwin, Robert, 219
Das Kapital, 222
David Taylor Model Basin, 272
Davy, Humphrey, 145
Defoe, Daniel, 304
Deism, 5, 223, 253, 284
dephlogisticated air (oxygen), 139
Descartes, Renee, 103
Dialogue between the Gout and Mr. Franklin,
    33
diving bell, 260
Divis, Procopius, 81
double hull, 264
double spectacles. *See* bifocals
Douglass William, 52
DuFay, Charles, 83
dysentery, 58
earthquake, 195-7
earth's core, 215
eclipse, 239
Edison, Thomas, 111
electric battery, 104
electric jack, 108
electrical stimulation, 99–102
electrochemistry, 145
electrodynamics, 108
electromagnet, 110
electron microscope, 48
electrons, 48, 81, 98, 110, 115-6 226
electrophore, 103
electroplating, 146
electrostatic generator, 82
electrotherapeutics, 106
elephant's tooth, 199
Elizabeth, Queen, 11
Ellis, Joseph, 12
empyema, 22, 49, 69
epidemic, 57, 58, 62
epidemiology, 59
Essay on Projects, 304
Essays to do Good, 304
ether, 39
Evans, Lewis, 200
evolution, 218
exercise, 36, 49, 77, 130, 155, 161-2, 166,
    175-8, 236
*Experimental Enquiry into the Properties of
    Blood*, 187
*Experiments and Observation on Different Kinds
    of Airs*, 132
Experiments and Observations on Electricity, 10,
    48, 88, 91, 92, 222, 244, 254-5
fabric, 111, 249, 250
Fahrenheit, Gabriel, 233

Faraday, Michael, 108, 109, 146
Fenno, John, 21
Ferdinand II (Duke of Tuscany), 233
fermentation, 4, 117, 121, 123-4, 129,
    153, 159, 179, 190, 310
fibrin, 188
fibrinogen, 186, 188
Finger, Stanley, 12
fireplace, 251-2, 284
Fitch, John, 279, 284-5, 289
fixed air (carbon dioxide), 4, 117, 124,
    129-31, 137, 138, 281
flatulence, 157
flea, 61
Fleming, Alexander, 68-9
Foley, Frederick, 31
Folger, Timothy, 259
fore-and-aft sail alignment, 262
fossils, 198-9, 201, 205, 217
Fothergill, John, 76, 92
four elements (Greek), 114, 119
four humors, 34, 159
Fourier, Joseph, 97
Franklin Court, 306
Franklin Institute, 306
Franklin stove, 252
Franklin, Abiah, 37
Franklin, Benjamin
    Chronology, 19-20
    Death Announcement, 21-22
    floating continents, 213
    plant-animal ecology, 131
    retirement from business, 84
    scientific progress, 148
    The oxygen-carbon dioxide cycle, 131
    theory of colds, 77
    theory of continent formation, 206-9
    theory of the earth, 201
    value of earthquakes, 211
    value of labor, 222
    vegetarianism, 155
Franklin, Deborah Reed, 14, 19, 64, 65,
    76, 130, 155, 169, 233, 268
Franklin, Francis (Franky) Folger, 51
Franklin, James, 19, 52, 239, 253, 304
Franklin, Jane (Mecom), 51, 53
Franklin, John, 29, 31, 37
Franklin, Josiah, 19, 37
Franklin, Peter, 251
Franklin, Sarah (Sally), 13, 19
Franklin, William, 19, 176, 230
Franklin, William Temple, 14, 15, 27, 65,
    230
Franklin-Folger Chart, 259
Friends of Franklin, 9, 320
Fulton, Robert, 288, 289
Galen, 159
Galilei, Galileo, 233
Galvani, Luigi, 103-6, 176

gaol fever, 61
Garrod, Alfred, 34
Gazette of the United States, 21
Genesis, 195
Gentleman's Magazine, 87
Geoffroy, Etienne-Francois, 126
geologic chronology, 204
geology, 197,199-205, 214-7 253, 261
George III, 97, 143
Germany, 61, 75, 83, 99, 125, 309
Glasgow Infirmary, 187
Godwin, William, 221
Goodman, Roy, 9, 185
Goodyear, Charles, 31
Gould, Stephen Jay, 295
gout, 33–36
Gray, Stephen, 83
Great Compromise, 28
Great Fire of 1666, 62
Grotto del Cane, 122
Guillotin, Joseph-Ignace, 297, 307
Gulf Stream, 137, 258, 259, 260
gunpowder, 20, 95, 96, 141-5, 160
Hadley, John, 39
Haiti, 67, 234
Hales, Stephan, 124
Hall, David, 84, 169
Hamilton, Hugh, 240
Harrison, John, 268
Harvard, 132, 195, 250
Harvey, William, 189
Hauksbee, Francis, 82
Hawaii, 269, 270
healthy living, 36
heart, 5, 31, 35, 54, 102, 106, 115, 173,
    179-80, 188-90
heat and light, 249
Heberden, William, 34, 53, 237
Helvetius, Mme, 42
Hewson, Polly see Stevenson, Polly
Hewson, William, 183-4
Hippocrates, 34
Historical Society of Pennsylvania, 206
History of Early Opinions Concerning Jesus
    Christ, 143
homeothermy, 179
Hooke, Robert, 44-48, 62, 123, 233-5
Hopkinson, Thomas, 85, 229, 286
hot-air balloon, 230
hoving-to, 266
How to secure Houses, &c. from
    LIGHTNING, 91
Howard, Luke, 241
Hunter, John, 53, 184
Hunter, William, 185
hurricane, 234
Husson Elixir, 34, 36
Hutton, James, 203, 204, 209, 214, 217
Huygens, Christian, 233

hygrometer, 234-5, 247
iatrochymistry, 125
immunity, 50-1
Impregnating Water with Fixed Air in order to communicate to it the particular Spirit and Virtue of Pymont Water, 130
inflammable air (hydrogen), 134
Ingenhousz, Jan, 101-3, 139-40, 160, 229, 230, 308
inoculation, 51-5
Introduction to a Plan for Benefiting the New Zealanders, 268
James I, 11
Jamestown, 62
Jay, John, 27
Jefferson, Thomas, 25, 27, 54, 143, 289, 291, 307
Jeffries, John, 230
Jenner, Edward, 53, 146
jet boats, 265
Johnson, Samuel, 64
Jones, John, 13-5, 20-6
Jones, John Paul, 14, 16, 264
Junto, 161, 162, 250
Kalm, Peter, 199
kidney stones. *See* stones, bladder
Kinnersley, Ebenezer, 85, 92
Koch, Pobert, 50
Kruse, Walter, 75
Lamarck, Jean-Baptiste, 172
Lamarckism, 172
Laplace, Pierre-Simon, 97
Lathrop, John, 323
Lavoisier, Antoine Laurent, 126-9, 136, 137-47, 227, 291, 297, 300, 307
lead poisoning, 12, 149–51
Leeuwenhoek. *See* Van Leeuwenhoek, Anton
Leon, Ponce de, 259
LeRoy, Jean-Baptiste, 290
LeRoy, Julien-David, 261
Leyden jar, 83, 84, 86, 103, 109, 160
light bulb, 111
lighter-than-air ships, 229
lightning, 4, 10-2 19, 79-97, 103, 106, 113-5, 142, 195, 225, 239, 253, 255, 300, 312
lightning at sea, 264
lightning rod, 10, 12, 19, 81-3, 87-95, 106, 110, 195, 255, 264
Lincoln, Abraham, 166
lines of magnetic force, 226
Lining, John, 159, 182, 236
lithosphere, 216
lithotomy, 38
Livingston, Robert, 289
lodestone, 82
Logan, James, 100
lost anchors, 262

Louis XIV, 65
Louis XVI, 20, 97, 141-4, 227, 297, 307
louse, 61
Luc, Jean du, 235
Lunar Society of Birmingham, 291
Lyell, Charles, 9, 197, 213-4, 217, 220
Madeira, 13, 64
Madison, James, 27
magnet, 82, 108-10, 116, 208-9, 216, 226
magnetic poles, 201
malaria, 63–66
Malthus, Robert, 218, 220-2
Margold, Marty, 256
Marly-le-ville, 89
Martin, David, 26
Marx, Karl, 222
Maryland, 10, 93, 197, 243, 244
Masons, 10
Massachusetts, 53
Massachusetts Magazine, 25
Mastodon, 199
matching grant, 301
Mather, Cotton, 52, 304
Mather, Samuel, 304
Mayor
    Boston, 28
    Philadelphia, 28
Mendeleyev, Dimitri, 146-8
Mesmer, Franz Antoine, 141, 255, 295-9
metabolism, 179–82
Methode de Nomenclature chimique, 142
Mexico, 307
miasma, 57
Michell, John, 211
microbes, 48, 50, 58, 61, 183
microscope, 47–48
Millikan, Robert, 81
Mitchell, John, 67
Moffatt, Thomas, 67
Monroe, James, 27
Montgolfier, Jacque-Etienne, 229-33
Morse, Samuel F. B., 110
mosquitoes, 63
Musschenbroek, Peter van, 83
Natural History, 202
Neave, Oliver, 215
New England Courant, 52
New Jersey, 57
New System of Chemical Philosophy, 146
New Zealand, 267, 268, 269
Newcomen, Thomas, 280-8
Newton, Isaac, 11, 85, 92, 115, 124, 128, 207, 211
Newton, Issac, 11
Nollet, Abbe, 84, 86, 89
Northwest Passage, 269
Novum Organum, 11
nutrition, 162

Observations concerning the Increase of Mankind and the Peopling of Countries, 221
ocean salinity, 253, 273–74
oil of vitriol, 227
oil on water, 275–78
On Franklin's Views of Ventilation, 122
On Thermometers, 233
opium, 64
Opticks, 11
Origin of the Species, 219
Ørsted, Hans Christian, 108
Oxford University, 18
oxygen. *See* dephlogisitcated air
Packard Humanities Institute, 10, 116
paddle wheels, 265
Paine. Thomas, 38
palsies, 4, 79, 99, 101
Papers of Benjamin Franklin, 10, 116
Paracelsus, 119, 125
parasite, 63
Parliament, 41
Passy, 25, 229, 297, 305
Pasteur, Louis, 50
Patent Act, 289
penicillin, 69
Pennsylvania fireplace, 159
Pennsylvania Gazette, 19, 25, 51, 52, 53, 58, 61, 91, 122, 196
Pennsylvania Hospital, 14, 163-4, 301-4
Periodic Table of Elements, 148
Peruvian bark, 64, 75
Petty, William, 263
Philadelphia Academy. *See* University of Pennsylvania
philosopher's stone, 120
phlogiston, 125-6 129, 136-43
phonetic alphabets, 219
Piccard, August, 230
piston, 280, 282
placebo-controlled clinical trial, 299
plate tectonics, 201, 205, 217
Poisson, Simon-Denis, 97
political parties, 28
polyp, 47
Poor Richard's Almanack, 14-5, 19, 64, 155
population growth, 221
Postal Service, U. S., 28
Postmaster General, 28
Price, Richard, 62
Price, Thomas, 195
Priestley, Joseph, 31, 81, 97, 110-1, 128-48, 231, 291
Princeton University, 100
Principles of Geology, 219
Pringle, John, 49, 61, 99-101, 130, 211, 271, 278
Progressive lenses, 26
propeller, 264

Proposals Relating to the Education of Youth in Pensilvania, 178
psychiatrists, 102
psychosomatic ailments, 297
Purfleet, 4, 79, 93-6, 129, 142
Pyrmont water, 130
quack, 35
Quaker, 92, 146-7, 161
rain gauge, 237
rainfall, 237
Ramsey, John, 165
rats, 61, 62
Rawley, Thomas, 25
Ray, Catherine, 172
Réaumur, René, 160, 233
red blood cell, 187
Redfield, William, 241
Redi, Francesco, 49
Reed, Wlater, 67
resinous electricity, 83, 97
Revolutionary War, 49, 97, 134, 137, 140, 141, 149, 165, 184, 227, 278, 284, 308
Richmann, Georg, 91
Ricketts, Howard, 61
Rickettsia, 61
Rittenhouse, David, 252
Robertson, John, 95
rockets, 231
Ross, John, 35
Ross, Ronald, 63, 67
Royal Society of London, 82, 85, 89, 104, 129, 184, 204, 211, 228, 240, 243, 267
Rozier, Jean Francis Pilatre de, 229
Rumsey, James, 279, 284, 287-9
Rumseyian Society, 279, 286
Rush, Benjamin, 67, 72-4, 130, 165
safety at sea, 261, 262
Sandwich Islands, 269
Sandwich, Earl of, 269
Savery, Thomas, 280
Scheele, Karl, 33, 128
schooners, 262
scurff, 49
sea anchor, 266
Second Continental Congress, 27, 137
seismic waves, 215
sentry box, 88-9
Seven Years' War, 307
Several methods of making salt-petre, 141
Sherman, Roger, 289
ship fever, 61
Ship leak, 262
single electric fluid, 98, 105
slave, 67
slavery, 308
Sloane, Hans, 200
Small, Alexander, 38, 122

smallpox, 51–56, 58
smoothing water, 275
Snow, John, 59
Some Account of the Pennsylvania Hospital, 303
Some Account of the Success of Inoculation for the Small-Pox in England and America, 53
Soulavie, Je3an Louis Giraud-, 207-17
sound waves, 215
South Pole, 269
space travel, 212, 231
Spencer, Archibold, 84
spontaneous generation, 49
St. Andrew's University, 18
St. Paul's Cathedral, 95
St. Petersburg, 91
Stahl, Georg Ernst, 125
Stamp Act, 20, 41, 129, 221, 284
steam engine, 266, 279, 280-90
steamboat, 279, 285-9, 311
steerable airships, 230
Stevenson, Margaret, 183
Stevenson, Mary (Polly), 183, 249, 279
Stiles, Ezra, 113-5
stones, bladder, 37–40
stratigraphic analysis, 205
Student's Elements of Geology, 197
sulfuric acid. *See* oil of vitriol
Suppositions and Conjectures on the Aurora Borealis, 225
Supreme Court, 28
swimming, 5, 173, 175, 178, 266
Swimming Hall of Fame, 175
Syng, Phillip, 85
Table of Affinities, 126
telephone, 111
The Art of Procuring Health and Long Life, 71
The Autobiography of Benjamin Franklin, 43
The Descent of Man, 220
The Enquirer, 221
The History and Present State of Electricity, with original Experiments, 97, 129
The History of the Corruptions of Christianity, 143
The Mathematical Principles of Natural Philosophy, 11
The Report of the Royal Commission, 295
The Sceptical Chymist, 123
Theory of the Earth, 202
thermometer, 159-60 233-6, 247, 259-60
thermometer conversion, 233
To All Captains and Commanders of American Armed Ships, 270
torpedo fish, 82
Traite elementaire de chimie, 143
transcontinental railroad, 290

transit of Mercury, 267
Tryon, Thomas, 155
tuberculosis, 50
Turgot, A. R. J., 97, 300
turkey, 88
Twain, Mark, 311
typhus, 61
Ultex lens, 26
Unitarian, 132, 140, 143
University of Pennsylvani, 53, 178
uric acid, 33
vaccine, 67
Van Helmont, Jan Baptista, 121-2
Van Leeuwenhoek, Anton, 47
Vassall, William, 52, 53
Vaughn, Benjamin, 149-52, 234, 287-8
vegetarianism. *See* Franklin, Benjamin, vegetarianism
Vesalius, Andreas, 175
viruses, 50, 51, 71, 75
vitreous electricity, 83, 97
volcanoes, 197
Volta, Alessandro, 4, 79, 103-8, 170
Voltaic pile, 104
Walpole, Horace, 35
War of Independence. *See* Revolutionary War
Washington, George, 27, 141, 164-5, 279, 284
watertight compartments, 263
Watson, William, 84-5, 89, 92, 95
Watt, James, 266, 279, 281-91
weather prediction, 239, 248, 253
Wedgwood, Josiah, 219, 291
Wegener, Alfred, 205
West, William, 92
Westminster Abbey, 237
wet nurses, 172
Whatley, George, 25
Whiston, William, 202, 207, 212
white blood cell, 187
White House, 96
Wilson, Benjamin, 95-6
Wilson, J. Tuzo, 205
Winthrop, John, 195
Withering, William, 291
Woodward, John, 201
Wyndham, William, 175
Yale, 10, 113, 116
yellow fever, 67–68
Zeppelin, Ferdinand von, 230
Zoll, Paul, 106
Zoonomia, 172, 219

From: "Stuart Green" <stuartgreenmd@yahoo.com>
To: "Benjamin Franklin" <dr_benjamin_franklin@yahoo.com>
Subject: **Disappointment and fallacious hope**

Dear Doctor Franklin:

I've just returned from Philadelphia. Upon arrival there, I went immediately to the excavation site at 4th and Arch Street. Imagine my dismay when I saw the completed National Constitution Center situated proudly over ground where I assumed you'd be found.

Somewhat reluctantly, I entered the building. The first exhibit occupies a curved hallway, with a large painted mural on one side and small display boxes on the other. The mural depicts life in Philadelphia of 1787. Life-size images of ordinary people—shoppers, merchants, and slaves—stroll along a busy street, going about their business. I visualized you in that picture, carried by sturdy prisoners to the State House in your French chaise.

The display cases built into the opposite wall, however, caused me great consternation. They contain items found during the Center's construction. While excavating, workmen made some astonishing discoveries. Digging alongside archeologists, they uncovered Philadelphia's first Native American artifacts, two revolutionary war muskets, numerous examples of colonial era pottery, clay pipes, knives and forks and spoons, and many coins as well.

The National Constitution Center's president recently proclaimed, "no one expected that the archaeologists would unearth so many extraordinary items, which help tell the story of an emerging nation." Most significantly, they dug up the remains of more than 90 individuals buried in an old cemetery. But what they *didn't* find distressed me the most—namely, you!

My worst fear, an erroneous assumption of your encaskment, has come to pass. Or has it? There remains one possible site Dr. Jones and your grandsons might have reached with a heavy cask on the night of your alleged demise: the park, just behind the Pennsylvania Statehouse (now called Independence Hall). Today, that land belongs to the federal government. Unfortunately for the both of us, Sir, the authorities won't likely issue a permit to dig for an oak cask containing the wine-soaked body of Benjamin Franklin!

Do you remember quoting Thomson's definition of "The Happy Man" in your 1755 *Poor Richard's Almanack*? Here it is: "A solid life, estranged to disappointment and fallacious hope, rich in content." My own existence, by this formula, is only half happy. I certainly enjoy a solid life (as a physician), and my travels around the world have been rich in content. Today, however, I'm no stranger to disappointment; I've clung to a fallacious hope of your revival.

It seems my speculation about your faked demise must remain just that—a conjecture, neither sustained nor refuted. Nevertheless, I've enjoyed our discourse, one-sided though it was. I've reviewed with you both the breadth and depth of your interests in science, medicine and technology, subjects that I too find fascinating.

Even if you never return, Dr. Franklin, I still enjoy rereading what you wrote in 1788 to Reverend John Lathrop, one time the pastor of Boston's Old North Church and later a Unitarian minister:

> I have been long impressed with the same sentiments you so well express, of the growing felicity of mankind from the improvements in philosophy, morals, politicks, and even the conveniences of common living by the invention and acquisition of new and useful utensils and instruments, that I have sometimes almost wished it had been my destiny to be born two or three centuries hence. For inventions of improvement are prolific, and beget more of their kind. The present progress is rapid. Many of great importance, now unthought of, will before that period be procured; and then I might not only enjoy their advantages, but have my curiosity satisfied in knowing what they are to be. I see a little absurdity in what I have just written, but it is to a friend who will wink and let it pass, while I mention one reason more for such a wish, which is that if the art of physic shall be improved in proportion with other arts, we may then be able to avoid diseases, and live as long as the patriarchs in Genesis, to which I suppose we should make little objection.

Now I must decide what to do with these numerous emails. I considered eliminating the entire sequence by simply selecting the *Franklin* folder on my computer's screen and then striking the "delete" key. However, another thought occurred to me: I could offer them to *Friends of Franklin*, to share them with others who, like me, are interested in your life and times. It's easy enough to do. I'll just attach the folder to an email to Roy Goodman, the organization's current president and off it goes as soon as I click the "Send" button.

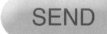